SCIENCE VS. RELIGION

SCIENCE VS. RELIGION

THE SEARCH FOR A RATIONAL APPROACH

Guido O. Perez

Algora Publishing
New York

Library of Congress Cataloging-in-Publication Data —

Perez, Guido O.
  Science or religion : a false dilemma / Guido O. Perez.
      pages cm
  Includes bibliographical references and index.
    ISBN 978-1-62894-105-0 (soft cover : alk. paper) — ISBN 978-1-62894-106-7 (hard
cover: alk. paper) — ISBN 978-1-62894-107-4 (ebook) 1. Religion and science. I. Title.
    BL240.3.P4725 2015
    201'.65—dc23
                                    2015009690

Printed in the United States

Table of Contents

Most people look to both science and religion to understand the nature of reality. They tend to entertain a mixture of views of the natural world when it comes to the origin and evolution of life and the universe and the emergence of mind and culture. Science embraces the naturalistic worldview, the view that nothing exists outside the natural world. But not all naturalists are atheists; some are agnostic or adopt a moderate version of naturalism referred to as methodological naturalism. By contrast, the major religions postulate the existence of a divine being that created everything that exists and interacts with his creation. This worldview, based on belief in the supernatural, says that beyond the natural world there is a spiritual reality that is not bound by physical laws. Not all who believe this are theists; some, like deists or pantheists, are non-theists (terms that we'll look into later).

The views of science and religion are very important in the lives of most people because they give them a sense that they understand the world and provide answers to many existential problems. Unfortunately, the debate between religion and science is becoming more acerbic each day. This book endeavors to engage in a civil and rational discussion of these important subjects.

Critics of religious fundamentalism blame institutional religion for many problems in the world. While it is true that much intolerance and violence has been fueled by religious differences, religion doesn't have a monopoly on evil actions. At the same time, religious institutions have sponsored artistic creations that enrich our culture and have provided assistance to the poor and the sick. Religious believers also may reject atheism out of fear that it could lead to moral nihilism; however, morality associated with divine commands sometimes breeds moral chaos.

For some people, religion provides purpose and meaning, and it can be a source of inner peace and happiness. Many nonbelievers, though, find that life can be meaningful in the absence of any imagined cosmic purpose. Others ask whether it is simply arrogance for humans to believe their lives should or could have a meaning beyond whatever impact they and their accomplishments have on the world around them. And anyone can experience a great sense of awe and wonder when contemplating the universe and the mysteries of life, without imputing the existence of a supreme being behind it all. Some nonbelievers consider that the exhilaration that comes from pondering the unanswerable questions is more authentic than being inspired by what appear to superstitions from the primitive past.

Both science and religion fall short of being able to account for every aspect of human experience. Neither science nor religion can prove or disprove the existence of an all-powerful, creator God. In addition, there is a wide spectrum of religious and scientific beliefs which various people find satisfying in different ways. These range from ontological naturalism and religious fundamentalism to religious naturalism and non-theistic religion.

This book describes the limitations of human knowledge and the impossibility of explaining everything. The author proposes a worldview based on scientific knowledge and logical reasoning, but he points out that trying to make either science or religion answer all our questions is a false endeavor. While science has its place in explaining many aspects of life, there are some questions that nag the human mind which can only be explored, not answered. It is the role of philosophy to go beyond science in this sphere, where there can be no empirical evidence or "proof."

The book presents a survey of the basic tenets of today's most widespread religions, and provides a detailed review of the scientific evidence that helps us understand the world. This review includes the most recent advances in evolutionary psychology, sociology, molecular biology and modern physics. The information presented raises important questions about the tenets of ontological naturalism and theism.

CHAPTER 1. INTRODUCTION

Throughout history many individuals have struggled to make sense of the world and to understand themselves. Primitive men explained their existence, as well as most natural phenomena, through animism. For primitive people, spirits or supernatural entities imbued all natural processes. As man began to congregate into societies and a division of labor arose, the priestly class flourished and religion became institutionalized. Early societies began to worship a pantheon of gods. During this period, they created major deities that explained human existence, as well as minor deities that held dominion over people's everyday lives.

During the period called the "Axial Age," 800 BCE to 200 CE, the spiritual foundations for many religions were laid simultaneously and independently in all the major emerging societies. The new religious systems reflected the changed economic and social conditions. In most of these societies polytheism gave way to monotheism. The explanation for our existence and the governance of our everyday lives fell under the purview of a single, omniscient, omnipotent, supernatural being. Today, the majority of the population of the planet believes in the existence of a single deity (as in the Jewish, Christian and Islamic religions).

As traditional religions grew some people began to question their faith and began to search for alternative explanations of the human condition and the origins of our existence based on reason and experience. Early Greek philosophers were among the first to try to understand the world in a systematic and logical manner. Other cultures also made contributions to early science. Many philosophers posed questions outside the framework of organized religion and discussed previously uncontested issues, such as free will and realism.

The foundations for the advancement of scientific knowledge were laid by the works of Galileo, Newton, Kepler, Darwin and others. During the age of

Enlightenment scientists added to the body of scientific knowledge and contributed to the establishment of scientific methods to explain natural phenomena. Since then, our understanding of the universe, as well as ourselves, has grown exponentially.

As soon as science began to explain the natural world a conflict began to emerge between science and religion. Copernicus suggested that the earth was not the center of the solar system and Galileo showed that he was right. It was not until 1822 that the Catholic Church formally acknowledged this new observation. In the nineteenth century the theory of evolution and Darwin's reply to Paley's argument from design was a source of conflict between science and religion. When Darwin published *The Descent of Man*, in 1871, the Pope denounced it, and many people claimed that the soul could not have evolved naturally.

As science advanced religion retreated to a higher ground. Creationism was embraced by religion to explain how we and the world came into being. Young-earth creationists maintain that the Genesis story is true and God created the world in six days around 4004 BCE, a date calculated from the Hebrew Scriptures. Old-earth creationists reinterpret the Genesis story so as to accommodate the standard geological chronology. Of course, every religion has its own creation story, of which at most one could be true.

Intelligent design theorists have suggested that many structures are so complex that they could not have been produced by evolution and the accumulation of small mutations. Behe claimed that several complex biological systems such as flagella and the blood clotting system could not have originated from small changes. There is no reason, though, to believe that a rudimentary organ, like a light-sensitive spot, would not be useful to the creature that has it. In spite of strong scientific evidence, Johnson argued that Darwinists have failed to produce the essential fossil evidence of intermediate species.

Science and religion differ from each other in many important ways. They have a different conception of the origin and the essential character of the universe. Science relies on experience and reasoning and eschews appeals to the supernatural. Religion is not primarily a kind of inquiry but a body of beliefs. Accordingly, it posits that the universe was created by a purposeful spiritual being and gave human beings a very special place. This being is concerned about our behavior and can be influenced by prayers or rituals. Commitment to this belief system in the absence of compelling evidence is considered a virtue. Unlike religion, theology is a form of inquiry. It accepts supernatural explanations, religious experiences and the authority of revealed texts.

Although most people claim that science and religion are incompatible with each other, there are those who affirm that they are compatible as long as God doesn't intervene in the world violating the laws of physics. Some argue that we are hybrid beings who exhibit both emotions and a rational mind, science studies the properties of physical things and religion addresses existential concerns. Gould has claimed that science and religion have distinct and non-overlapping domains. Science is concerned with the realm of facts, whereas religion with the realm of values. Davies has argued that the belief in God is largely a matter of taste and Swinburne advanced the notion that the hypothesis that there is a God explains everything. He treats religious experience as a real phenomenon. Many people believe in various aspects of both science and religion but don't mix their religious beliefs with their scientific knowledge; they prefer to keep them in separate compartments.

The intellectual debate in the West has been dominated by either the rationalistic and scientific outlook of the Greek philosophers or that of the Romantics who placed emotion, intuition and religious beliefs before reason. Thus we tend to embrace either science (reason) or religion (faith, religious experiences) to understand the world in which we live (Flanagan, 2003). The former is based on the worldview of naturalism, the belief that the natural world is all there is, and the latter on supernaturalism, the view that, in addition to nature, there is a deity and a spiritual realm that is not bound by physical laws. Many people embrace science and rationalism over religion. Nevertheless, religion is still prevalent; about five in six people in the world have a religious affiliation. Those without religious affiliation comprise the third most numerous group after Christians and Muslims.

The debate between those with and those without religious beliefs is confounded by the imprecision of the terms used in the articulation of their respective views. For example, strictly speaking atheists are people who reject theism, the conception of God upheld by the three major monotheistic religions. Some people, though, believe that the term atheism should be reserved for individuals who reject the idea that God exists. Negative atheists are people who are simply devoid of religious beliefs; whereas positive atheists don't believe in the existence of any deity

There are strong and weak forms of theism and atheism; some make ontological and others epistemological claims. For example, the strong or positive form of ontological theism holds that God exists, whereas its weaker or negative form claims that it is possible that God exists. The strong form of epistemological theism affirms that one knows that it is the case that God exists and the weak form claims that one knows that it is possible that God exists. One may be an ontological theist but not an epistemological theist; fideists, for example, assert that God exists but rely on faith rather than reason to justify their beliefs. Likewise, there are atheists who hold that is not the case that God exists without having enough epistemological justification, meaning that one cannot prove a negative statement. The last two are weaker positions; they are forms of belief without knowledge, that is, without having adequate justification.

There are important limitations to either the scientific or the religious worldview. Scientists don't have the capacity to observe reality directly, they can only create models of the world that then can be rejected or corroborated by empirical means. These models are only approximations of reality and are sometimes changed as new information becomes available. We still don't have a successful "theory of everything" or an adequate description of the subatomic world. Naturalism is sometimes metaphysically speculative and its proponents may be influenced by their own wishes and temperament.

Another shortcoming of science is that it doesn't give answers to existential issues or provides us with meaning and purpose in our lives. Secular humanism, though, puts emphasis in this life here and now, the quest for value and the good life and the institution of social justice and equality for all. It also affirms that values grow out of human experience and can be examined critically. Secular humanists hold that the planet Earth should be viewed as our common abode and each individual has an obligation to preserve the environment and to promote some form of population restraint.

Philosophy is often considered a branch of science. Continental philosophy took a phenomenological and existentialist turn. Most of their followers affirmed that the universe is pointless and that human freedom would be compromised by the existence of a deity. Analytic philosophy endeavored to differentiate between analytical and synthetic propositions by conceptual analysis. Many philosophers of language have questioned the logical consistency and coherency of religious claims. A widely held view is that philosophy helps to formulate questions and clarify concepts but it offers no definitive answers about the nature of reality. Because many philosophers have not kept up with modern developments in science and scientists don't have a definitive description of reality, some people believe that physics and metaphysics should be integrated into a single discipline.

The comforting answers provided by religion are based on faith and revelation, not on evidence. There is a striking lack of evidence to support the traditional religious claims and Scripture often doesn't agree with current scientific knowledge. The hiddenness of God and the problem of evil argue against some of the tenets of theism. In spite of these limitations, many people have difficulty in living with the fact that scientific hypotheses are tentative and don't solve most existential problems. They cling to the certitudes offered by the religious worldview that was prevalent in their culture of origin. For many people knowledge of God is not something waiting to be discovered by rational means; it is the result of a transcendent experience, the immediate and personal experience of being connected to some divine order.

In addition to the limitations of the traditional scientific and religious view of reality, there are numerous competing ideas within each worldview. Scientific views range from the extreme position of the "new atheists" to more moderate positions. Ontological naturalists are usually positive atheists and physicalists or materialists. Some of them are reductionists, whereas others are emergent naturalists who claim that some entities and processes can't be explained by lower levels of complexity. Methodological naturalists resort to science to understand reality, but don't deny the existence of supernatural entities and processes, and don't claim that everything is physical. Agnosticism holds that the evidence for and against the existence of God is such that one cannot affirm or deny religious beliefs; agnostics are usually engaged in a continuous quest to find the truth. Agnosticism also comes in ontological and epistemological variants. Not all individuals without religious beliefs belong to the previous categories; there are also skeptics, humanists and religious naturalists, people that find spiritual comfort in the beauty and wonders of nature.

There is also a wide spectrum of religious views and different belief systems. Some theists are exclusivists who claim that their beliefs are the only true ones; salvation is denied to those from other religions or to individuals who had no access to the revealed "truth." Inclusivists allow that other religions may have some truths and pluralists affirm that other religions represent equally valid truths or ways of life to gain access to the divine reality. There are also believers that are non-theists, like deists, pantheists and those who embrace process theology or liberal religions.

Chapter 2. The Two Major Worldviews

Supernaturalism

The two leading views about the nature of reality are those of supernaturalism and naturalism. Supernaturalists believe in the existence of an agency that can influence the course of natural events. They include theists, individuals who believe in a deity that has several specific attributes and created everything that exists (Dennet, 2006) and non-theists. Naturalists believe that nothing exists outside the natural world and embrace the scientific method to understand reality. Ontological naturalists deny that God exists and hold that the universe is composed of only physical entities. Methodological naturalists prefer to focus on human affairs or nature and avoid the issues of physicalism and God's existence. Thus not all naturalists are atheists. Likewise, not all supernaturalists are theists; there are deists, pantheists and those that follow process theology.

Theism is embraced by Judaism, Christianity and Islam, the three Abrahamic religions that had their origin in the ancient religion of the Jews. Theism affirms that God is an omniscient, omnipotent, omnipresent and beneficent being that is eternal, self-existent, created the universe but is separate from it and intervenes in the world. The theistic conception of God is only supported by faith and claims of revelation; many of the characteristics attributed to God contradict each other, thus undermining the credibility of claims of "revealed truth." Believers dismiss this problem, saying that God's attributes are hidden from us; God doesn't unequivocally disclose his nature or clearly state his wishes to all humanity. The problem of evil, especially natural evil, suggests that God is either not beneficent or that he is not in control of the fate of his creation; this undermines the premise of God's fundamental characteristics.

Critics note that both the Hebrew Bible and the New Testament contradict science. For instance, they point out that as far as anyone can see, it is just a self-serving story — albeit, a very appealing one — that imperfect human beings were created following God's image or that we enjoy a special status in nature. By contrast, the theory of evolution holds that all living species evolved from a primitive ancestor, each species was not created separately. Some Christians accept the evidence of evolution but still add the groundless assertion that it is directed by God. Believers maintain that their beliefs are supported by other people's reports of miracles and religious experiences. Miracles and religious experiences, though, contradict the laws of physics and cannot be proven. And as any legal expert can tell us, witnesses who claim to know what happened even in every-day events often tell conflicting stories about it. This is not reliable evidence.

Theists claim that God is the cause of all events that science cannot explain. Nonetheless, science keeps explaining more of the mysteries of nature and there are many scientific theories that may one day provide an answer to the remaining questions. Theists believe that without God life would be meaningless and we would not have a sense of morality, whereas non-religious people ask why humans would believe that life has meaning at all. Or they infuse it with their own private, personal meaning and that morality results from an evolutionary adaptation to communal life.

Sin and salvation, heaven and hell, miracles and the power of faith and prayer are other beliefs embraced by Christians and other theists. Revelation is the concept that truth and knowledge can be disclosed, or "revealed," that some form of truth or knowledge can be imparted to humans through communication with a deity or other supernatural entity.

For Christians, this means accepting as reliable any information about God contained in the New Testament (a collection of writings said to embody knowledge that was revealed to men). Christian theism requires embracing additional beliefs. The Christian doctrine of the Trinity holds that there is one God who is also "three persons." Incarnation is the claim that Jesus is God represented in a flesh-and-blood man. Resurrection is the idea that after being killed, Jesus was brought to life again. At the heart of traditional Christianity are the rituals of initiation or Baptism and the Eucharist, a re-enacting of the ancient tradition of sacrifice, as those who partake are told that they are eating the flesh of their god and drinking his blood.

There are several non-theistic conceptions of God that make fewer assertions that have to be accepted on faith, but there is no way to examine even these conceptions in a rational way. These include the view that God created all there is and set up the laws of nature, but no longer interacts with his creation (deism); the belief that God is everything and everything is God, the totality constituting a divine unity (pantheism); and the view that the natural world resides within the divine being (panentheism). Process theology holds that God, just as the rest of the natural order, is in the process of becoming. While such beliefs are free from the dogmatism of traditional religions, they are still mental constructs that cannot be tested or supported by any evidence.

Hinduism and Buddhism, the two other large religions, had their roots in Asia. The most common thread of what we consider Hinduism is reverence for the Vedic Scripture, a rich collection of works that includes the highly philosophical Upanishads. According to Advaita Vedanta, the world is an illusion and behind that which we experience is the formless, impersonal reality of Brahman. Enlightenment or Moksha is the result of overcoming the dualism of Brahman and the Atman or individual soul; it ends the cycle of birth and rebirth. Hindus may be monist or polytheistic, but most believe that a trinity of Brahma, Shiva and Vishnu is the cardinal, supreme manifestation of Brahman.

Buddhism originated in northern India from the teachings of Gautama Sakyamuni who came to be known as the Buddha. The Buddha taught the four Noble Truths, a strategy to stop suffering by extinguishing cravings. The Dharma is the teachings of the Buddha, the path to attaining enlightenment and the Sangha is the noble community, those advanced in the understanding of the tradition. Buddhism is non-theistic and underscores the absence of self or the Atman and claims that everything is impermanent. Like Hinduism, it includes a belief in reincarnation, but denies the existence or importance of Brahman. The main schools of Buddhism are Theravada Buddhism, prevalent in Sri Lanka and Southeast Asia, Mahayana Buddhism practiced in East Asia and Tibetan Buddhism.

The standard orthodox interpretation is that Theravada Buddhism is atheistic. In its original form, it did not support a belief in the type of god associated with the Abrahamic religions. Nevertheless, some experts have claimed that belief in devas, impermanent gods borrowed from Hinduism, suggest that Buddhism is a theistic religion. Jainism, one of the large Indian religions has no teachings about the existence of God. Its founder, Mahavira opposed ritualistic killing of animals, the caste system and the idea that the individuality of the soul is absorbed into Brahman. Confucianism is also an atheistic religion. Although Confucius was an agnostic humanist, his views were submerged in a cultural context of superstition. The only supernatural beings are good men that were capable of achieving perfection.

Naturalism

Most scientists are naturalists, they believe that all that exists is part of nature and that knowledge about these entities comes from scientific inquiry. Methodological naturalists restrict themselves to the use of science to understand reality; ontological naturalists go far beyond and assert that everything is physical and that supernatural entities don't exist. Some ontological naturalists are positive atheists who hold a materialistic and reductionistic view (Stenger, 2003). Others argue for emergent naturalism, the view that certain complex phenomena can't be explained by lower levels of complexity.

Most ontological naturalists affirm that our universe is eternal and uncreated, perhaps one of many in a multiverse unlimited in time and space; some believe that it had a beginning, whereas others assert it is cyclical. They believe that science alone gives us a good understanding of the workings of the universe. For them quantum theory explains the subatomic world and is the theoretical background of modern chemistry, an important tool to understand molecular biology. They claim that life and mind originated by natural processes and that most life processes rely on the working of molecular machines, different proteins that carry out different cellular functions.

Other non-believers include agnostics who claim that we cannot prove or disprove the existence of God, secular humanists, individuals who prefer to concentrate more in world affairs and religious naturalists who reject the supernatural but find spiritual comfort in nature. Religious naturalists believe that their spirituality represents the constellation of emotional and mental experiences associated with contemplating the different aspects of the natural world. They include the awareness of our place in the cosmos, the interconnection among all life forms and the awareness that we are all made of the very stuff that formed during cosmic evolution.

The following chapters review the different ways that the two major worldviews attempt to answer the most important questions about the nature of the world including ourselves. The main questions include the origin and expansion of the universe, the appearance and evolution of life, the tree of life and the relationship between mind and brain. Subsequent chapters explore how the religious and the scientific worldviews explain human nature, including the issues of free will, morality, politics, economy and existential concerns. The scientific view is developed in great detail because of its explanatory power, its rapid growth in the last few years and the fact that very few theists are acquainted with the naturalistic worldview and need to improve their scientific literacy.

CHAPTER 3. THE ORIGIN OF THE UNIVERSE

The Beginning

Most cultures have tried unsuccessfully to explain how and why the universe originated. In ancient Greece, Epicurus held that the universe is all there is and that everything can be explained by the chance combination of atomic elements. Aquinas held that God, a self-existing or necessary entity, must have been the first efficient cause of everything that exists. He rejected the idea of an infinite series of regressive causes. Thus he held that if everything were contingent nothing would exist; God is the only entity that could have caused the universe to exist. Some people, however, claim that the universe itself may also be past-eternal, necessary and self-existing. Others claim that the entropy problem requires that the age of the universe is finite. Some philosophers hold that it makes no sense to talk about anything, natural or supernatural, as existing necessarily. Others claim that the origin of the universe doesn't have an explanation; they reject the principle of sufficient reason and claim that the existence of the world is a brute fact.

Many religions postulate that the universe had a beginning and that God created it from nothing (ex-nihilo) or from unformed chaos. There are also scientific theories that suggest that the universe could have originated ex-nihilo from a random quantum fluctuation of vacuum. These theories assert that even if the energy of the quantum field is zero, it is never really zero due to the fluctuations. Critics, however, argue that empty space may not be the same as nothing or nonexistence. The raw materials and the laws postulated to explain the spontaneous formation of the universe could not have arisen from nothing.

Many believers assert that the existence of anything is a miracle caused by God that violates the first law of thermodynamics (the law of conservation of energy). According to Hawking, though, matter and energy are interchangeable and the existence of matter/energy doesn't represent a miracle. This is because the total matter/energy of the universe is zero; the negative gravitational energy exactly cancels the positive energy represented by matter.

Religions offer many myths in which a divine entity creates the universe. In our culture we adopted the Jewish creation myth which in turn was borrowed from those of Mesopotamia. The biblical account of the early author "J," however, is different from that of author "P." It is unclear if God created the universe out of nothing or from preexisting matter, or if God exists forever "in" time (everlastingness) or "outside" time (timelessness). Many religions claim that God exists "outside" time. It is difficult, though, to understand how an entity that is outside time can perform an act that is necessarily temporal. In the Babylonian creation epic (Enuma Elish), God and the cosmic matter were coeternal and the creation activity unfolds throughout six days in the same order found in Genesis.

There is strong scientific evidence that suggests that our universe began abruptly from a state of infinite density and temperature (the big-bang). The big-bang theory has been upheld by many religious leaders who claim that it supports the creation story described in Genesis. We really don't understand, however, what happened before, or at the moment of the big-bang or what was what "banged."

The big-bang theory leads to a singularity in which all physical theories fall apart. To understand this theory, physicists need to develop a theoretical model that unites quantum mechanics with relativity. In the "no boundary" proposal of Hawking and Hartle, the universe is self-contained, without a beginning or end. Loop quantum gravity holds that the universe developed from a primordial state described by quantum gravity, a theory that applies quantum mechanics to general relativity. Other recent models of the origin of the universe are based on M-theory and inflation (Smolin, 2006; Hawking and Mlodinow, 2010).

M-theory and eternal inflation predict the formation of multiple universes (multiverse). M-theory is an extension of string theory, a theory that posits that all particles may be themselves made of tiny loops of vibrating energy. M-theory theory postulates the existence of two-dimensional membranes and higher dimensional branes, it allows for the formation of each with different laws depending on how their multiple dimensions are curled up.

According to the Hawking and Mlodinow model the origin of the universe was a quantum event that could be described by Feynman sum over histories.

They suggested that because there are many universes, a sum over them should be taken to get the probability of the universe that we observe. This model doesn't include singularities; it postulates a cyclic universe without inflation and having no beginning and no end in time. In the cyclic model of Steinhardt and Turok adjacent branes collide and separate cyclically, each collision being a collapse and each separation a new big-bang. It doesn't postulate a creation event. Smolin's model proposes that multiple universes are generated through black holes.

Inflationary theory explains how repulsive gravity blew up the universe in the first few fractions of a second after the big-bang. According to inflationary theory, a special source of energy, called a scalar field, was responsible for the rapid and sustained expansion of the universe. Because of instability it decayed very fast to allow inflation to end. The inflationary model is supported by recent studies of the cosmic microwave background (CMB) radiation. The observed temperature and density fluctuations in the CMB may have originated from quantum fluctuations in the scalar field. Recent unconfirmed observations with the Biceps-2 telescope suggest the existence of gravitational waves, relic ripples in the fabric of space and time from the inflation era. The inflationary model, though, doesn't explain what happened just before the big-bang.

A model of the universe, possibly without a beginning, is that of eternal inflation. It proposes an eternal creation process resulting in an infinite number of "island universes" each with its local laws of physics (Greene, 2011; Vilenkin, 2006). Such a theory, although difficult to prove, would explain the improbability of our universe, the observation that the fundamental physical constants of nature seem to be fine-tuned for the development of life; our universe is what it is because we are here. Alternative explanations for the fine-tuning of the universal constants include the possibility that the constants are interdependent, are changing or that true random events do happen.

Cosmic Expansion

Since the big-bang most cosmic events have been found to have a natural explanation. The concept that the universe is static was shattered in the early twentieth century when Hubble found that galaxies are drifting away at a velocity proportional to their distance from us. The discovery was made because the galaxies' light is red- shifted, that is, stretched to longer wavelengths because of the Doppler Effect. Far from being static the universe may be growing in size.

By imagining the expansion running back, astronomers deduced that the universe originated 13.7 billion years ago, at the time of the big-bang. The age of the universe was confirmed by measurements of the Hubble constant, the most recent data from the Planck telescope suggests that the universe is a little older, about 13.8 billion years old. In the 1980's theorists added the phase of rapid growth called inflation, and more recently it has been discovered that cosmic expansion began to accelerate again several billion years ago under the influence of a form of repulsive gravity dubbed "dark energy".

All there is left of the big-bang at the present time is the fairly uniform CMB radiation of 2.7 degrees above absolute zero discovered by Penzias and Wilson in 1965. The existence of this relic of the big-bang has been confirmed by NASA's Cosmic Background Explorer (COBE), the Wilkinson Microwave Anisotropy Probe (WMAP) and by the Planck telescope. The last one had greater precision and confirmed the presence of areas of lumpiness in the CMB.

After the big-bang space itself expanded, it was not an explosion where objects are projected from each other. Physicists posit that the primordial plasma started to cool off in the first one hundredth of a second after the big-bang, allowing the formation of protons and neutrons from quarks. After three minutes, as the simmering universe cooled to about a billion degrees, hydrogen and helium nuclei were formed (stage of nucleosynthesis, coupling of matter and energy). About three hundred and eighty thousand years after the big-bang, as the universe continued to cool, free electrons were captured by nuclei forming the first atoms (decoupling of matter and energy). Prior to this era the universe was opaque because photons were captured by the dense plasma of charged particles.

Matter and antimatter should have been created in equal amounts. Sakharov proposed that as the universe cooled, a slight asymmetry developed, and as the matter-antimatter annihilated, a very small amount of matter remained. From that small remnant, stars, planets and people like us ultimately formed. As of this writing there are no experimental studies definitely explaining the origin of the matter-antimatter asymmetry.

The big-bang created mostly hydrogen atoms, small amounts of helium and traces of lithium. As the universe cooled, lumps of matter started to clump together forming the first stars. These stars were much heavier than our sun and gravity pulled inward the atoms of matter to a common center. Nuclear fusion in the core of these early stars produced the first twenty-six elements of the periodic table. When the core of these stars filled up with heavier elements gravity took over and the stars collapsed inward, they subsequently exploded as supernovae leaving their residue as a black hole. The explosion seeded the space with carbon, oxygen, nitrogen, phosphorus and sulfur, the elements found in our bodies, and silicon, iron, magnesium, aluminum and calcium, the main components of rocky planets. The scattered debris of old exploded stars, under the organizing force of gravity, produced new stars and galaxies.

For galaxies to have formed there must have been some kind of unevenness in the universe at a very early stage. As mentioned above, this unevenness was detected by the recent precise observations of the WMAP satellite and the Planck telescope. Early galaxies were smaller than more recent ones suggesting that the latter were formed by the agglomeration of smaller galaxies. One theory of galaxy formation posits that highly energetic quasars preceded the formation of galaxies. Astronomers have suggested that quasars, distant, bright and active sources of radiation, are powered by giant black holes that occupy the center of most galaxies.

Stars have a life cycle, they are born, evolve and die when they run out of energy from fusion (Lang, 2013). If a star's mass is less than a certain limit (Chandrasekar limit, or one and a half times the mass of the sun) it becomes a brown or white dwarf. If it is above that limit, gravitational collapse produces either a pulsar (a rapidly spinning neutron star that emits narrow beams of electromagnetic radiation that rotate like giant lighthouse beams as the pulsar spins) or a black hole (an infinitely dense, dimensionless point from which not even light can escape).

About 4.5 billion years ago a nebular cloud, halfway out from the Milky Way galaxy center, initiated our Solar System. The observed orbital uniformity in our Solar System suggests that the planets and moons coalesced from the same rotating disc of dust and gas at more or less the same time. Our Solar System occupies a small section in one of the arms of the Milky Way galaxy. The Milky Way is a spiral galaxy that looks like a flat disc with a bulge in the middle. There are two to four hundred billion stars in our galaxy, each rotating around a massive black hole about once every two hundred and fifty million years. To get a grasp on the size of the universe we need to realize that the Milky Way together with the Andromeda galaxy and about thirty other small galaxies form a small cluster referred to as the Local Group. The latter is in turn a tiny piece of the Virgo Supercluster, which is about 100 billion times larger than the Milky Way.

The position of Jupiter in the Solar System defined our planet's mass and its distance from the sun. As a result, the primitive Earth enjoyed the ideal conditions for the development and evolution of life. Our planet is right in the middle of the "Goldilocks" zone in its nearly circular orbit around the Sun. It is close enough to the Sun to have a rocky composition (the intense solar wind pushed the lighter gaseous elements to form the giant planets), and far enough from the Sun to hold water in liquid form. We don't know if similar conditions exist in other parts of the universe.

The moon probably originated from the collision of Theia, a smaller would-be planet, with the primitive Earth. Since its formation, the Moon has been moving away from Earth, the explanation of this phenomenon may be found in the law of conservation of angular momentum. The Earth's angle of axis rotation with respect to the axis of the orbital plane is stabilized by the Moon, and the giant planets keep most destructive comets and asteroids away from us. The rotation of the Earth's iron-rich core creates the magnetic field that protects us from damaging solar wind and the ozone layer shields us from noxious ultraviolet radiation.

The Earth's geological time is divided into two eons: the Pre-Cambrian, going from the formation of the planet to five hundred and fifty million years ago, and the Phanerozoic, extending from that time to the present. The Pre-Cambrian eon is divided into the Hadean, the Archean and the Proterozoic; the Phanerozoic is divided into the Paleozoic (ending 251 million years ago), the Mesozoic (ending 66 million years ago) and the Cenozoic (from 66 million years ago to the present time) eras. The Paleozoic includes the Cambrian, Ordovician, Silurian, Devonian, Carboniferous and Permian periods. The Mesozoic encompasses the Triassic, Jurassic and Cretaceous periods. The Cenozoic is divided into the Tertiary (including the Paleocene, Eocene, Oligocene, Miocene and Pliocene epochs) and the Quaternary (including the Pleistocene and Holocene epochs).

Early in the history of the Earth, global-scale volcanism separated water and other volatiles from the basaltic magma to create the first oceans and the atmosphere (Lenton, 2011). The planet's other possible sources of water include the impacts of icy comets or asteroids. As the Earth cooled the silicon-rich magma pushed its way toward the surface giving rise to the continents. Life may have emerged from inanimate matter as early as 3.5 billion years ago. The appearance of photosynthetic organisms was responsible for the Great Oxidation Event. Oxygen is the ideal electron acceptor; its release from photosynthesis oxidized iron to form the rusty red mineral hematite. The next billion years were relatively quiet until the end of the Proterozoic, when the Earth underwent severe snowball-hothouse cycles. The recent past has been marked by the effects of plate tectonics molding the continents and by the progressive rise of the terrestrial biosphere.

During the last 2.3 million years our planet has cycled through several periods of cold and dry weather commonly known as the ice ages. The succession of glacial-interglacial periods took place during the Pleistocene (Paleolithic or Old Stone Age in archeological terms). At present the earth is in an interglacial period, the Holocene. The changes in climate produced substantial changes in plant, animal and human populations. The cause of the cycles of heating and cooling is not well understood. The Milankovich effect posits that they result from the net effect of three events: changes in the shape of the Earth's orbit around the Sun, precession or obliquity of the Earth's axis of spin and the tilt or wobble of Earth's spin.

The earth behaves like a giant organism in which all parts, including the biosphere, are interconnected. The heat from the Sun causes evaporation of water from the oceans. In the cooler layers of the atmosphere water vapor condenses forming clouds. When the amount of water in the clouds exceeds a certain limit, water comes down to Earth in the form of life- sustaining rain. The energy released by electrical storms reacts with atmospheric nitrogen to form nitrates, a well-known and important fertilizer. Forest fires triggered by lightning help to renew the forests and return their rich minerals to the soil.

Ocean currents tend to stabilize global temperatures. The reaction of hot magma with cold water in the bottom of the oceans creates a surplus of chemical nutrients that stimulate the formation of algal blooms which in turn feeds marine life. Phytoplankton is one of the most important sources of oxygen in the atmosphere. Dust from the Sahara desert, which is rich in dead marine organisms, often blows to the West, helping to nourish the soils in the Amazonian forest. The rich minerals that originate in the large mountain ranges are washed by rivers to the lowlands thus improving the quality of the soils. Human activity also contributes to changes in the planet. We create large amounts of greenhouse gases, release large amounts of sulfur and other pollutants into the atmosphere, cause deforestation, deplete aquifers and contribute to the extinction of many species.

The energy from the Sun should be available for a few billion years, enough time is left for our climate to change, plate tectonics to keep shuffling the continents, which in turn, may cause life to evolve. Eventually, the Sun will get progressively hotter and will become a red giant vaporizing our planet by engulfing our atmosphere or creating intense solar winds. In the distant future we have to worry about desertification caused by the progressive increases in the Sun's temperature, asteroid impacts, collision with other galaxies, eruption of giant volcanoes and the return of the ice ages. In the next few years, though, the biggest threats to our planet are man-made: global warming caused by greenhouse gases, pandemics, overpopulation, scarcity of water and other resources, nuclear war, and terrorism associated with nuclear, cyber or biotechnology.

Twenty billion years from now the Milky Way may collide with the Andromeda galaxy. Eventually, the universe as we know will also disappear. The fate of the universe will depend on the struggle between its rate of expansion and the pull of gravity, which is primarily dependent on the density of matter. Matter ruled the universe in the past but its influence at the present time is smaller. The future of the universe depends on whether "dark energy" will continue to increase causing further cosmic expansion, or will decrease leading to a contraction of the universe.

Chapter 4. The Evolution of Life

Origin of Life

Most religions explain the origin of life as an act of divine intervention. For scientists the origin of life from inanimate matter must have been a natural phenomenon. The way it happened remains unexplained, but scientists have developed two main theories: "the metabolism first" theory and the "replication first" theory. Panspermia claims that life came from outer space; it does not explain the origin of life elsewhere (Deamer, 2011).

The most popular theory for the origin of life posits the formation of a macromolecule capable of self-assembly into liquid phase droplets and self-reproduction. RNA is the most likely candidate macromolecule because it is both a catalyst and a carrier of genetic information. RNA is unstable and eventually DNA took over the function of storing genetic information. Other scientists have embraced the theory that metabolism came first, life could not be sustained without a continuous flow of energy; metabolism must have preceded informatics and the RNA world (DeDuve, 2005; Lane, 2005).

For life as we know to arise there is a need of liquid water, carbon-based molecules, an energy source and time. It is unclear how often all of these conditions obtain in the universe and how likely it is for life to arise when it occurs. There is a philosophical disagreement between scientists who claim that life was bound to arise given the prevailing physical-chemical conditions in the early universe and those that that hold that the origin of life was an improbable event. The former tend to support the presence of life in other parts of the universe and the claim that the fundamental constants of nature appear to be fine-tuned for life to emerge. Those who claim that the origin of life was an improbable event suggest that we may be alone in the universe and that if we would rerun "the tape of life" we would get a different result each time.

The transition from inanimate matter to life was relatively fast, but a sharp or clear-cut transition may not have taken place. All life forms consist of one or many cells and all life processes are chemical in nature, determined ultimately by physics, the behavior of atoms. Life is difficult to define, the difference between inanimate objects and living organisms is that the latter reproduce, grow, respond to stimuli and perform biochemical functions carried out by thousands of proteins which are themselves coded by genes. Proteins convert chemical energy into useful work enabling cells to move, transport and perform all the activities that keep them alive.

Life appears to violate the second law of thermodynamics because ordered entities cannot arise from less ordered systems. Life forms, though, are open systems that receive energy from the outside in the form of food and are capable to maintain a temporary energy imbalance and evade decay to energy equilibrium. Life is a continuous fight against entropy. The controlled flow of protons across the mitochondrial membranes drives the generation of energy and order in complex life forms. The waste products produced during metabolism increase the overall disorder outside the organism so overall entropy in the world increases.

The catabolism of energy-rich carbon molecules involves the transfer of electrons from these molecules to oxygen, a highly electronegative compound, ultimately ending with the synthesis of ATP, the energy currency of the cell. The energy-rich carbon molecules are synthesized in green plants by the process of photosynthesis using the energy of the sun and producing oxygen as a byproduct. Without plants we wouldn't exist. They provide the oxygen that we breathe, the materials to make our clothing, the medicines to cure disease and the food from which we obtain our nutrition. In addition, most of the energy that we use in the developed world comes from the combustion of fossil fuels that originate from dead trees.

The Theory of Evolution

In the nineteenth century Darwin advanced the theory of evolution by natural selection. The theory of evolution provides a compelling picture of how complex life forms originated from the first cells. It is supported by overwhelming evidence from a variety of fields including biogeography, paleontology, geology, embryology, biochemistry, developmental biology, genetics and molecular biology. This theory is the best explanation we have for the origin of the wide variety of species found in the living world.

Darwin had concerns about negative reactions from the Church, but his supporter Huxley didn't share such reservations. He engaged in heated debates about evolution with Bishop Wilberforce of Oxford. Wallace, the co-discoverer of evolution didn't achieve as much notoriety as Darwin probably because he didn't provide enough experimental data to support the theory and because of his interest in spiritualism. The term "survival of the fittest" was not introduced by Darwin in the first edition of Origins. It was introduced in the fifth edition after a suggestion by Spencer who felt that competition between individuals should drive the species forward and create a better society made up of healthy and fit individuals. Spencer became the main proponent of Social Darwinism, and suggested that governments should not support poor people because this would encourage further breeding of unfit individuals. This theory gained the support of many industrialists, particularly in the United States, in spite of the reservations of Moore and others who claimed that Social Darwinism was an example of the naturalistic fallacy; that is deriving "ought" from "is".

Galton took the notion of Social Darwinism a step further and defended the implementation of Eugenics. This theory was based on the idea that the State should control human reproduction in order to rid itself of weak, feeble-minded or sick individuals. This movement gained many supporters in the first three decades of the twentieth century and was the reason why thousands of people were forcibly sterilized. It was the precursor of the Nazi regime's policies in the 1930s and 1940s, which put to death millions of people in the Holocaust in the name of racial purity.

In 1871, Darwin published *The Descent of Man and Selection in Relation to Sex*. In this book he demonstrated that certain human features such as sympathy and reasoning are observed to some degree in animals. In the second part of the book he introduced the concept of sexual selection, the suggestion that individuals may be selected purely on their ability to find mates. In another book he compared the facial expressions between different species and between different cultures. His main argument was that people of different racial origin use the same expressions to demonstrate the same emotions. Among the similarities that he observed between humans and animals was the rise in the hair in the back of the neck caused by fear, the showing of teeth induced by anger and the closing of teeth tightly when concentrating. This work was not accepted by many social scientists; they rejected the Blank Slate view of human nature and concentrated more on environmental factors. It took a long time before the impact of these books was appreciated.

Darwin died in 1882; at the time of his death his theory of evolution by natural selection had dropped out of favor, in part because the mechanism of heredity was not yet known. In Germany, Ernst Haeckel elaborated his own version of Darwinism, which added to the discussion throughout Western Europe. Some of his theories, like the view that ontogeny recapitulates phylogeny, and that humans can be separated into ten separate races, were rejected by many of his fellow scientists. Haeckel also suggested that there were distinct stages of evolution that were driven by a Lamarckian model, the inheritance of characteristics that were acquired during the individual's lifetime.

In the first part of the twentieth century, the new field of population genetics helped to integrate Mendelian genetics with Darwinism and this, in turn, led to what it became known as the Modern Synthesis (Maynard-Smith and Szathmary, 1995; Dawkins, 2009). This theory explained how the accumulation of small genetic changes and the reordering of the genes over many generations explained evolution. According to this theory natural selection is the main mechanism of evolutionary change, evolution is a gradual process and macroscale phenomena can be explained by the microscopic process of gene variation and natural selection.

There have been several challenges to the Modern Synthesis. The primacy of natural selection in the evolutionary process has been questioned by some scientists; genetic drift and self-organization may also play a role in evolution. The view that small gene variations may explain the history of life has also been challenged. It is still unclear how and where evolution operates and how evolutionary forces shape the genome. In addition, others like Gould have proposed that evolutionary changes may have not been smooth and gradual (phyletic gradualism) but rather jerky and abrupt (punctuated equilibrium).

Both Marx and Freud were influenced by Darwin's ideas. The former dedicated the German edition of *Das Kapital* to Darwin, and the latter suggested that human emotions were originally adaptive and that the libido was related to sexual selection and the mating instincts. According to Freud, we are born with biological forces that propel us toward objects of desire and self-preservation.

Darwinism was criticized by Agassiz, who believed in racial differences and subscribed to "creation science." James, a disciple of Agassiz, studied how consciousness functions to help humans adapt to their environment. In contrast to Darwin he suggested that experience follows rather than precedes emotional expression. Boas' cultural relativism challenged Darwinism, insisting that we should reject theories of human nature based on evolution but look instead to environmental influences. In the nature/nurture controversy, he became a supporter of the blank slate theory. His work also countered the racist stereotypes about people from other cultures.

Mead also disputed the idea that our emotional expressions are universal and arose via natural selection. Some of her field work, however, lost credibility and was criticized by others due to errors in data collection. The social sciences' conflicts with Darwinism arose because of their rejection of racist assumptions which included the supremacy of white Western societies. Some of these debates re-emerged later with the ascent of sociobiology.

Darwin's work had a major impact on ethology, the field of inquiry into animal behavior. The books written by Tinbergen, Lorenz and Morris became best-sellers and helped to popularize the study of animal behavior. They also supported the development of sociobiology, behavioral ecology and evolutionary psychology. Their conclusions, though, were based on the characteristics of European society at that time and offered a male-centered view of evolution. Comparative psychology also drew on the work of Darwin. This discipline emphasized the learning ability of the animal under controlled experimental conditions. It branched into the study of animal cognition and eventually into behaviorism, propelled by the studies of Watson and Skinner. To behaviorists behavior was learned during an individual's lifetime and not by natural or sexual selection. Skinner called the form of learning that he was studying "operant conditioning" because the animal operates on the environment and learns to increase positive responses and to avoid negative ones. He believed that language develops through conditioning with parents rewarding the correct use of language. This theory was opposed by Chomsky who suggested that babies are born with an innate language-acquisition device.

Evolutionary biologists took ideas from Darwin about natural and sexual selection and integrated them into modern explanations of behavior by shifting emphasis away from the level of the species to the level of the individual or the gene. This led to the development of the new fields of sociobiology and behavioral ecology. Wynne-Edwards pioneered the use of the group selection theory to explain many aspects of animal social behavior; cooperative and reproductive behavior ultimately evolved to help the survival of the group. His theory helped explain why the animals engage in self-sacrificing behavior if a major part of natural selection is to survive and reproduce. Nevertheless, this theory implies that over many generations selfish individuals would outbreed the selfless ones. Maynard-Smith and Williams drew on the original suggestion by Darwin that when animals do cooperate, it is likely to be between close relatives. This suggests that when animals behave altruistically to their kin they were actually acting on behalf of their own genes. These authors shifted the emphasis to the individual and genetic level of selection.

Self-sacrificing behavior outside of the close relative relationship was explained by the work of Hamilton who specialized in the social behavior of ants. Hamilton realized that the worker ants could pass copies of their genes indirectly by working to help raise their younger sisters rather than breeding themselves. This process became known as kin selection and has now been expanded into the inclusive fitness (the number of copies of our genes that are passed into future generations) theory. Trivers introduce the concept of reciprocal altruism to explain altruistic behavior between unrelated individuals.

The Darwinian explanations of pro-social behavior also gained support from the new field of Sociobiology and from the writings of Wilson and Dawkins. Both authors espoused a gene-centered view of selection. Wilson created a heated controversy by suggesting that human behavior can be explained by use of Darwinian principles and claiming that ethicists and social scientists may be influenced by sociobiology. Dawkins message was more moderate and he didn't delve into the issue of ethical behavior.

There were many scientific and political criticisms of sociobiology. Many people rejected Wilson's idea that the patriarchal nature of many societies can be traced back to foraging strategies between the sexes. His view that war mongering behavior had played an important role in our evolution was also challenged. His critics accused him of propagating right-wing political ideas. He was labelled a "biological determinist," attempting to provide a genetic justification of the status quo. Dawkins was accused of "adaptionism," the tendency to see all features of animals as having arisen by Darwinian selection to help their ancestors pass on their genes. Critics pointed out that some structures, like feathers, were used originally for insulation and much later were modified to achieve powered flight. Sociobiologists have defended themselves by rejecting genetic determinism and claiming that behavior requires the interaction of genetic and environmental factors.

Behavioral ecology was also influenced by Darwinism. It is based on empirical studies and places more emphasis on the flexibility of human behavior that sociobiology. Like the latter, they were accused by critics of making a link between foraging efficiency and "inclusive fitness." More recently, another field of study derived from Darwinism, evolutionary psychology, has captured the imagination of both scientists and the general public.

Evolutionary psychologists followed the ideas of sociobiology and cognitive psychology. One of the early proponents of the influence of mother–infant relationship on the emotional health of the individual was the British psychologist John Bowlby. He became convinced that the ethological approach to animals could be applied to human behavioral development.

When he read the work of Lorentz on imprinting he immediately recognized the importance of the influence of the quality of the infant's attachment to the caregiver on the emotional health of the individual. Trivers explained the reason why males compete for females as a case of parental investment. Because females invest more effort into raising offspring they are more discerning when it comes to the mating game. The work of Tooby and Cosmides is considered the foundation of evolutionary psychology. They rejected the blank slate theory and argued that the evolution of the mind by Darwinian mechanisms helped to cope with the challenges facing hunter-gatherer societies. They proposed that the mind consists of many modules that evolved to solve specific problems that our ancestors faced on a regular basis. They used the reverse engineering method, studying a particular trait to infer the reasons for that design.

The studies of Buss and Pinker also support the tenets of evolutionary psychology. Buss showed that males are attracted to female features associated with fertility, whereas females place more emphasis on the male's ability to provide resources. Pinker argued that the development of language has all the hallmarks of Darwinian adaptation, whereas Ridley and Miller asserted that language evolved through Darwin' mechanism of sexual selection. Critics of evolutionary psychology have argued against the notion that the mind was adapted by natural selection in the Pleistocene epoch and that it is made up of many independent modules and that can be understood by reverse engineering.

In genetic terms, evolution consists of changes in the organism's hereditary material (DNA). The structure of DNA was not established until the middle of the twentieth century. Mapping of the human genome at the beginning of the third millennium was considered to be the greatest achievement of molecular biology (Riddley, 1999). The original dogma of molecular biology was that DNA directs the synthesis of RNA and the latter directs the synthesis of proteins. Nevertheless, now we know that there are viruses that convert RNA into DNA and that many simple organisms carry more genes than humans do, suggesting an important role for regulatory systems that switch genes on and off (Carroll, 2005).

Recent studies have shown that only 1.8 percent of the DNA is transcribed into proteins. New genes may evolve by duplication followed by divergence and other mechanisms, duplicate gene copies may mutate into useless pseudogenes. Most of the remaining non-coding DNA is involved in gene regulation or evolved from invasive viruses and freeloaders. The former includes segments that don't code for proteins but for RNA fragments that have regulatory functions (Kazazian, 2011). The DNA in the genes has to be uploaded as strings of RNA. Recent studies indicate that whether or not the information in the gene is uploaded depends on how the DNA is packaged around proteins. When DNA is methylated it is tightly packed and silent, whereas when protein spools are acetylated it packs lightly and the DNA is actively read.

The vertebrate genome exhibits evidence of invasion by viral DNA or RNA. Retroviral genomes, the best understood viral insertions, account for about eight percent of human DNA. These insertions are inherited along with the rest of the host's DNA sequences. Although these sequences are usually defective, some may affect the host by assuming the control of host transcription or influencing the expression of adjacent host genes. The reactivated virus may reemerge as an infectious agent and infect the host or other species.

The source of evolutionary variation is introduced by errors in DNA replication, mutations induced by physical or chemical agents and the reshuffling of genes during meiosis. The changes that are "selected" and passed down to future generations are those that help an individual's probability of surviving and leaving offspring. Thus genetic variants arise randomly in populations but natural selection is not random. Chance and determinism work together to propel evolution towards increasing complexity. Of interest, chance and determinism also appear to work together to create the ordered universe that we observe.

Creationism and Intelligent Design

A large percent of the population in the United States doesn't believe that evolution by natural selection took place. Some follow the bible literally, whereas others claim that living organisms had to be designed by an intelligent being. Nevertheless, "creationism" and the "intelligent design" theories are not scientific theories as some of its proponents claim; they are contradicted by the theory of evolution.

In the beginning of the twentieth century the biblical creation story became the standard alternative to evolution. In 1925 a biology teacher was prosecuted for teaching evolution in his school. The so called "Monkey Trial" resulted in a public relation victory for those supporting evolution. Subsequently, a group of creationists developed the intelligent design theory which claimed that there is evidence of design in nature and that this evidence supported the existence of a creator. The theory was refuted by most scientists and its proponents never published their views in peer-reviewed journals.

The intelligent design proponents created the Discovery Institute to support their agenda. One of its members published a book asserting that chance cannot generate order. Nevertheless, as mentioned before, natural selection is not due to chance alone and entropy in nature can either increase or decrease. Living systems are open systems that use sources of outside energy to maintain order. In another highly publicized trial a court in Pennsylvania ruled that intelligent design was motivated by religion and presenting it in science classes was unconstitutional, the judge also concluded that intelligent design is not a scientific theory.

Some scientists have claimed that natural selection cannot account for the complexity of life. Most scientists agree that mechanisms other than natural selection, like genetic drift and self-organization, could play a role in evolution. Kaufmann has postulated that self-organization plays an important role in the origin and evolution of life. It is unclear, though, if self-organization represents a new holistic law in nature. With the use of computers it is possible to create cellular automata that have the ability to reproduce themselves. Wolfram has suggested that this may the beginning of a new science in which the universe behaves like a digital computer; he sees no need to postulate the existence of supernatural agents.

Intelligent design proponents, however, have attempted to prove that the existence of what they called "complex specified information" is evidence of a supernatural designer. Nonetheless, complex structures, such as the double spiral pattern widely found in nature, could arise from natural causes. Proponents of intelligent design claim that some biological systems show signs of pre-existing purpose or plan. Many scientists, though, argue that it has been shown many times that biological systems co-opt primitive systems and adopt them for a new biological function. For example, the hormone insulin that regulates blood sugar was originally a growth factor in unicellular organisms. In addition, there are many structures, such as the eye or the low back, that show evidence of what appears as "bad" design. The reason these structures are the way they are is their evolutionary history.

There are theistic evolutionists, including religious leaders, who believe that evolution took place but assert that God is in some way involved in the process and that the changes don't involve the human soul. Some theistic evolutionists believe think that God is still directing evolution, others claim that God set up the conditions for evolution but did not play a guiding role. They hold that evolution follows a divinely created law that tends to build more complexity. Theistic evolutionists point to a collection of examples of what is called convergent evolution, the discovery that some structures evolved numerous times along separate pathways but with the same goal in mind. The majority of evolutionary biologists accept that convergence often occurs, but there is no reason to believe that it occurs because of divine laws.

Evolutionary Developmental Biology

All cells in the body carry the same DNA, it is now well established that development, from a single cell or zygote, to the mature organism with different tissues and organs depends on the time and location when the genes are switched on and off. It turns out that the human genome is not a blueprint but it is more like a script that can be read several ways. For an embryo to develop into an adult organism requires an amazingly complex series of events in which the genome is transcribed dynamically in time and space. We now understand better the main conserved genetic pathways involved in setting up cell lineages and determining embryonic patterning. Specific cells within the early mammalian embryo, called embryonic stem cells, have the capacity to generate all somatic cell lineages plus the germline. In vivo this property, called pluripotency, lasts only a few days, but stem cells can be maintained indefinitely in tissue culture. This information has been a key factor to guide current research designed to turn adult cells into pluripotency.

Determining the genome-wide binding sites of lineage-specific transcription factors and integrating the information with genome-wide transcriptional profiling continues to provide information about the mechanisms involved in pluripotency and lineage differentiation. The new field of epigenetics explains how environmental stimuli such as diet or chemicals may reprogram our genes. It is now apparent that epigenetic modifications of DNA and its associated histone proteins provide another level of control by modulating the accessibility of DNA for transcription. Epigenetic mechanisms are well known to play a role in X chromosome inactivation (the process by which an X chromosome in female mammals becomes genetically inactive), and genomic imprinting (the process by which certain genes are differently expressed when inherited from mother or father). Another level of control takes place at the posttranscriptional level by means of short RNA sequences or microRNAs that bind messenger RNA or block translation.

A key event in the post-implantation development of the embryo is the formation of the primary germ layers, ectoderm, mesoderm and endoderm, during gastrulation. Cells allocated to these cell layers exhibit a progressive restriction in lineage potential. The specification of lineage progenitors is accompanied by the appropriate switches in the activity of the genome. For example, the formation of germ cell progenitors entails the activation of germline-specific genes and concurrent suppression of somatic cell genes. Superimposed on these transcriptional events are epigenetic modulators of gene activity and the regulatory activity of microRNAs.

Multicellular organisms can mold groups of cells into functionally specialized tissues and organs, a process called morphogenesis. To understand how cells organize into tissues requires precise knowledge of the cell-signaling pathways in which extracellular ligands bind to receptors and transduce signals to the interior of the cells. Organogenesis begins with the formation of the organ primordium, which is made up of the progenitors of the particular cell type and the establishment of the vascular nervous and lymphatic supply.

CHAPTER 5. THE TREE OF LIFE

According to evolutionary theory, all living organisms, including humans, may have descended from a single cell ancestor called the Last Universal Common Ancestor (LUCA). This early ancestor evolved into the prokaryotes (Bacteria and Archaea), and the eukaryotes. Because of lateral gene transfer the relationship between these three groups of organisms has been difficult to establish. Prokaryotes are unicellular organisms that lack a nucleus and membrane-bound organelles. Archaea differ from bacteria in the composition of their cell wall and plasma membrane and their habitat, which is often adapted to extreme environments. Fossils of bacteria similar to today's cyanobacteria have been found in rocks (stromatolites) from Western Australia that are 3.3 to 3.5 billion years old.

Prokaryotes used different mechanisms to convert precursors into energy-rich molecules. Autotrophic organisms that resemble today's cyanobacteria probably had the capacity to synthesize carbon-based molecules from water and carbon dioxide, releasing oxygen as a byproduct, utilizing the energy provided by the sun (photosynthesis). Some bacteria developed the capacity to extract hydrogen from hydrogen sulfide leaving elemental sulfur as a waste product. Eukaryotic cells have a membrane-bound nucleus containing DNA and membrane-bound organelles like mitochondria and chloroplasts. They appeared more than two billion years ago, it was an event that may have enabled the evolution of multicellular organisms. Heterotrophic eukaryotes obtained their energy from the catabolism of carbon-containing molecules in their mitochondria; these molecules were made in plants' chloroplasts by the process of photosynthesis.

Eukaryotes may have originated when two prokaryotes merged to create an endosymbiont. The organisms containing organelles responsible for energy production (mitochondria) originated when cells adopted prokaryotes capable to utilize the chemical energy of glucose. Organisms containing photosynthetic organelles (chloroplasts) probably originated when cyanobacteria that possessed a two-step photosynthetic mechanism were adopted by cells that already contained mitochondria. Early in the Proterozoic era (2–2.5 billion years ago) photosynthetic organisms began the progressive oxygenation of the atmosphere. This process, which eventually led to the formation of the ozone layer of the atmosphere, led to the proliferation of organisms capable to carry out aerobic metabolism. Aerobic metabolism is a more efficient source of energy than anaerobic metabolism, a factor that may have contributed to the development of multicellular life.

Prokaryotes, though, have not disappeared from the face of the Earth, bacteria are currently found in large numbers in the soil, in the oceans and even the most extreme habitats. In nature bacteria are responsible for the fixation of nitrogen by plants and for the phenomenon of putrefaction. Industrial uses of bacteria include sewage treatment, fermentation of food and the removal of oil spills. Of interest, human beings are colonized by a vast number of bacterial species. We have about ten times more bacteria than cells in our body. They occupy multiple sites, especially the skin, the vagina and the gastrointestinal tract. Collectively they are referred to as the human microbiota. Most of these organisms are not pathogenic and they may contribute to the maintenance of our health. Changes in the bacterial flora, especially in the gut, have been associated with inflammatory bowel disease, immune disorders and the overgrowth of pathogenic species.

Many bacteria are pathogenic and are known to invade and kill other organisms. Infectious diseases have been present since the dawn of humanity and have been responsible for epidemics that caused the death of millions of people. Tuberculosis is one of the oldest diseases affecting humanity; tuberculous bone disease has been found in Egyptian mummies. The disease reached epidemic proportions during the industrial revolution and still has not been eradicated. Plague's deadly epidemic potential has been well documented throughout history. The Justinian pandemic killed millions of people in Africa, Asia Minor and the Mediterranean basin. During the Middle Ages a second pandemic killed as much as a quarter of the affected population in Europe and became known as the Black Death.

A major step in the evolution of life on earth was the appearance of multicellular organisms (metazoans). Single cells cannot develop much complexity because of size limitations, to increase their mass they would require provision of unattainable flows of nutrients through the cell membrane. The first clear-cut evidence of multicellular life does not appear until about six hundred million years ago. Complex fossil algae specimens and molecular evidence suggests that metazoans could have existed earlier. It is important to point out that a haphazard accumulation of independent cells held together by adhesion molecules or symbiotic colonies formed by two or more organisms without central control (e.g., sponges) does not constitute a true multicellular organism. True multicellular organisms arose from single cells when they acquired a genetic program that enabled them to develop into daughter cells with different structures and functions.

There are many theories that explain the increase in the size and complexity of multicellular organisms. These changes may have been related to improved genetic programs, aerobic metabolism or to the evolutionary advantages of sexual reproduction. Asexual reproduction results in the progressive accumulation of DNA copying errors and vulnerability to environmental influences. Sexual reproduction by shuffling genes thoroughly with each generation creates a novel collection of genes. It helped organisms survive under adverse conditions but created the need for sharing genes, competing for mates and taking care of the offspring.

In organisms that reproduce sexually both sperm and eggs pass on one half of the genes in their nucleus to their offspring but only the egg passes mitochondria to the next generation. This is because the two sets of mitochondria compete with each other and the egg started out larger and with more organelles and resources than the sperm. Thus the egg carried all the resources necessary for providing for the offspring, whereas the sperm became slim and movable in order to compete more effectively with others and increase its chances of fertilizing an egg.

An upsurge in odd-looking metazoans appeared in the fossil record five hundred and seventy-five million years ago during the Ediacaran period. Ediacaran organisms may or may not have been the precursors of the large number of organisms that evolved during the Cambrian, the first period of the Paleozoic era. Early Cambrian organisms were followed by those exhibiting a great increase in size and diversity. The appearance of mineralized tissues, triggered by changes in the chemical composition of the oceans, may have influenced the balance between predators and prey.

During the Cambrian there was a gradual build-up of multicellular organisms over eighty million years, rather than a sudden explosion. After about three billion years of evolution in which there was nothing other than single-celled organisms, the Cambrian seas were swarming with multicellular organisms. The largest fossil deposits from this period were found in the Burgess Shale in British Columbia. The ancestors of the five major phyla, mollusks, arthropods, chordates, nematodes and annelids appeared during this time.

Metazoans are classified according to the number of germ layers: two in diploblasts (radiata) and three in triploblasts (bilateria).The diploblasts include the sponges, cnidarians and ctenophores. Their tissues originate from two layers: ectoderm and endoderm. The once radial symmetric animals eventually acquired an elongated shape with clear front and back, top and bottom and bilateral symmetry. These bilaterally symmetrical animals had a third layer or mesoderm between the endoderm and the ectoderm.

Bilaterally symmetrical animals split into Protostomia and Deuterostomia depending on the fate of the blastopore and the type of coelom formation. The blastopore is a small opening in the mass of cells (blastula) that develops shortly after fertilization. In Protostomia this opening becomes the mouth whereas in Deuterostomia it forms the anus, the mouth develops at the other end of the blastula. In the Deuterostomia the fluid-filled body cavity or coelom forms from the endoderm rather than the mesoderm.

Protostomia consists of two groups: Ecdysozoans and the Lophotrochozoans. The process of exoskeleton molting known as ecdysis is characteristic of the Ecdysozoans. The arthropods are the most important group of Ecdysozoans. Lophotrochozoans include mollusks, annelids and brachiopods. They derive their name from a tentacular feeding organ (lophophore) and a type of ciliated larva known as the trochophore.

Deuterostomia include the Echinoderms (starfish and sea urchins), Hemichordata and Chordata. These are our closest relatives among living animals. Hemichordata include organisms such as the acorn worms and the pterobranchs. The body plan of Chordata consists of two sections: a head with pharyngeal perforations (gill-slits) and a posterior segmented unit. The nerve chord and the notochord run along the back and the digestive system is found in the belly. The vertebrates, a subdivision of the phylum Chordata can be traced back to the lancelet and amphioxus, the precursors of jawless fishes. The central nervous system arose from the neural tube in vertebrates, the brain originated from bulges in the anterior part of the neural tube, whereas the posterior part became the spinal cord. The vertebrate design has an obvious advantage: the protection of the central nervous system by bony structures.

Fishes proliferated during the Devonian (mid Paleozoic). One of the largest fossil deposits from this period was found in Scotland. Louis Agassiz, the famous Swiss naturalist, wrote a large treatise on fossil fishes between 1833 and 1844. The hagfish and the lampreys are the only jawless vertebrates alive today. Before the origin of jaws, vertebrates were limited in what they could eat. Once this innovation appeared in the Silurian there was a rapid proliferation of jawed fish or gnathostomes. Initially, the now extinct Placoderms were dominant. They gave way to cartilaginous (rays and sharks) and later to the bony fishes. There are two groups of bony fishes, the "lobe-finned fish" and the "ray-finned fish." The final step in the evolution of "ray-finned fish" was the development of teleosts. They comprise most of the species found today.

A major transition in the history of life on earth was the movement of animals from water to land. Plants, mainly algae and lichens, and arthropods such as horseshoe crabs, were among the first living organisms that invaded the land. The first fossils of insects appear during the Silurian, about four hundred million years ago. Insects exhibit six legs, a head with compound eyes, thorax, abdomen and a chitin exoskeleton. They invented powered flight and proliferated to produce millions of species.

The step by step transition from fish to tetrapods, four legged land animals, is well documented in the fossil record. A recent expedition to the Arctic discovered Tiktaalik, a fossil that represents an intermediate stage between the lobe-finned aquatic Eusthenopteron and the footed Ichthyostega. Recent genome sequence studies, though, show that lungfish, not coelacanths, are the closest living relatives of tetrapods. These early tetrapods could have invaded terrestrial habitats and even climb trees as some fish continue to do today. To live on land they had to develop mechanisms to deal with gravity, to avoid dehydration and to extract oxygen from the air. In the modern phylogenetic or cladistic classification, which asserts that grouping of species should be strictly genealogical, amphibians are considered to be tetrapods that need to stay near an aquatic environment in order to reproduce.

During the Carboniferous most tetrapods branched into the monophyletic group of amniotes. Their complete break from the water was aided by the development of the amniotic egg. This egg has a hard shell to protect from evaporation and a system of inner membranes that surround the embryo, the food supply and the waste products. The development of scaly, waterproof skins and the production of near solid urine helped these animals to conserve water.

The amniotes are divided into three separate clades based on the number of fenestrae in their skulls. The Synapsids had only one fenestra; they were the ancestors of mammals. The Diapsids, with two fenestrae, included most reptiles, dinosaurs, and birds. The Anapsids, with no fenestrae, formed a group that included the turtles. The original crown-clade classification divided amniotes into two groups, the Synapsids (or ancestor of mammals) and the Sauropsids (that evolved into reptiles, and subsequently, into birds). Like all mammals, we descended from a fishlike ancestor; human embryos exhibit pharyngeal pouches (which become gill slits in fish) and a long tail about five weeks after conception.

Two hundred and forty-five million years ago, at the end of the Paleozoic, there was a catastrophic mass extinction (the Permian extinction). Seventy-five percent of reptiles and amphibians and fifty percent of marine families disappeared during a relatively short period. The cause of this mass extinction is unknown; it may have been the result of climate changes associated with volcanic eruptions.

The early Mesozoic witnessed the rapid proliferation of the dinosaurs. Dinosaurs are divided into two groups, the Saurischidian (with lizard-type hips, which include the Theropods), and the Ornithischian (those with bird-like hips). The transition from dinosaurs to birds during the Mesozoic Era was first suggested by the discovery of the feathered Archaeopterix fossil in 1860. Many transitional bird fossils and feathered dinosaurs have been described recently that fill the gap between Theropod dinosaurs and birds. Feathers probably did not evolve for flight but were originally used for insulation.

The lineages of synapsids and true reptiles diverged during the Early Carboniferous Period. The primitive synapsids evolved into pelycosaurs (dimetrodon) and therapsids that dominated the Late Permian landscape. The next grade up is the group called the cynodonts, which survived the Permian extinction at the end of the Paleozoic. The more advanced cynodonts from the Late Triassic looked more like modern mammals. They had a mammalian jaw joint, all non-dentary jaw bones were lost except the articular and quadrate bones which became incorporated in the chain of little bones that transmit sound from the eardrum to the inner ear. This change gave mammals the ability to discriminate sounds with much higher frequencies than those detected by birds and reptiles. It led to the development of a strong jaw and teeth.

The first mammals probably emerged from cynodonts in the Late Triassic, at the same time of the early dinosaurs. During this period mammals were small, nocturnal and looked like the modern shrew. At the end of the Mesozoic, about sixty five million years ago, another mass extinction wiped out the non-avian dinosaurs as well as most large land animals. The cause of the extinction may have been the climate changes produced by the impact of a large asteroid near the Yucatan peninsula. Birds and mammals, being endothermic, proliferated and occupied the ecological niches that became available after this catastrophic event. They co-evolved with flowering plants (angiosperms) and other pollinators.

Early mammals (monotremes) laid eggs like their amniote ancestors. Pouched marsupials and placental mammals evolved later. These animals were already evolving by the time that most non-avian dinosaurs disappeared. After rodents and bats, the third largest group of placental animals is that of the ungulates or hoofed mammals. Even-toed hoofed animals (Artiodactylia) are the most abundant ungulates in the planet. They were the ancestors of hippos and those mammals that returned to the oceans such as dolphins and whales. Carnivores have a complete fossil record that goes back to the entire Cenozoic. Dogs appeared in the Eocene and cats during the Early Oligocene. Humans, of course, are also placental animals.

The order of primates cannot be definitely characterized by a single or even multiple traits. Most of them became arboreal probably to take advantage of the newly formed ecological niche. Life in the trees was associated with two important evolutionary adaptations: the divergent thumb that permitted better grasp of tree branches; and the development of tricolor vision that allowed these animals to distinguish raw from ripe fruits. The latter took place when one of the old opsin genes was duplicated.

Traditionally, primates are divided into prosimians, such as lemurs and tarsiers, and anthropoids. The latter are divided into new and old world monkeys and hominoids. The hominoids include the lesser apes, the great apes and humans. The latest classification lumps together all great apes and humans as hominids and the species that appeared after our common ancestor with the chimpanzees are referred to as hominins. Molecular dating techniques, suggest that apes diverged from other primates seventeen to twenty five million years ago, the orangutan ancestors diverged from other apes about thirteen to sixteen million years ago, the gorilla ancestors diverged from the ancestors of chimpanzees, bonobos and humans seven to nine million years ago and those of chimpanzees and bonobos diverged from human ancestors five to seven million years ago.

The Human Lineage

Our first known ancestors, including *Sahelanthropus tchadensis*, *Ororrin tugenensis* and *Ardipithecus ramidus*, lived in Africa five to seven million years ago. The oldest fossils of the advanced genus *Australopithecus*, *A. anamensis*, were found in rock beds 3.9–4.2 million years old. The best known of the early australopithecines is *A. afarensis*, a female fossil ("Lucy") that displays a mix of human and apelike traits. It was found in beds about 3.2 million years old. More complete remains of *A. afarensis* have been found in Hadar, Ethiopia and Laetoli, Tanzania. The males were much larger than the females, they had small brains and large jaws, and their knees showed clear evidence that they walked upright. Bipedalism may have been a successful adaptation to cover long distances in the cold, dry savannas found in East Africa after the formation of the Rift Valley. Others have suggested that this change was present in our ancestors before they left the woodlands. Upright posture improved visibility and body cooling and liberated the hands for manual labor (Diamond, 1992).

Between 2.5 and 2.8 million years ago, two major branches evolved: the robusts (strong) and the early humans classified in the genus *Homo*. The earliest fossils thought to belong to our own genus, *Homo habilis*, meaning "handyman," have been discovered in beds about 2.8 million years old. The use of rudimentary stone tools (Olduwan tools) is associated with most species of *Homo*. This ancestor of modern man had a larger brain volume of six to eight hundred cubic centimeters which may have been sustained by a high quality diet that included meat. Hominins stayed in Africa until the emergence of the genus *Homo*.

Members of the genus *Homo* left Africa at least three times and invaded Asia and Europe. The first migration took place appeared about 1.9 million years ago. *H. erectus* (ergaster) fossils have been discovered in Georgia and Java (1.8 million years old), Beijing (half a million years old) and perhaps the Island of Flores. The second migration may have taken place across the Strait of Gibraltar into Europe; *H. antecessor* and *H. heidelbergensis* may be the earliest members of this group.

Neanderthals (*H. neanderthalensis*) probably evolved in Europe from *H. heidelbergensis*. The oldest skeletons are about one hundred and seventy-five thousand years old. Neanderthals reached their peak in the last interglacial, a hundred and twenty thousand years ago. They had large brains, comparable to ours, and a robust skeleton. They were hunters, used primitive stone tools and practiced ceremonial burials. Some investigators considered them as a population of archaic *H. sapiens*, whereas most recent studies argue for a separate species affiliation, *H. neanderthalensis*. They disappeared at the peak of the last ice age, about thirty thousand years ago. The origin of the Denisovans remains found in a Siberian cave and dated between thirty and fifty thousand years has not been clearly established.

H. erectus had a brain two-thirds the size of modern humans' and the skeleton had most of the hallmarks of modern man. The species probably lost most body hair in order to permit sweat glands to develop and to avoid skin infestations. *H. erectus* lasted until about thirty thousand years ago. More sophisticated stone axes and tools (Acheulian tools) and the first evidence of the use of fire comes from this period.

It was not until about two hundred thousand years ago when *H. sapiens* (modern humans) evolved in Africa. These hominins started to show large brains and physical characteristics of modern man. The majority of modern humans remained in Africa; they are the ancestors of native Africans. Behaviorally modern humans, though, emerged much later.

During the early part of the twentieth century, most experts believed that the transformation of archaic *H. sapiens* to modern humans was a worldwide event. Nevertheless, studies of the inheritance of the Y chromosome and mitochondrial DNA support the theory that the common ancestor of all men living today came from Africa and spread from there to the rest of the world. This third migration out of Africa began about sixty thousand years ago; it became the ancestral population that through many generations spread to colonize the rest of the world. It is unclear the role *H. sapiens* may have played in the demise of other hominins.

Mitochondrial DNA is a circular double stranded molecule that codes for thirteen oxidative phosphorylation components and several types of RNA. When the sperm and the egg fuse, all the sperm's mitochondria are destroyed, leaving the fertilized egg with only the mitochondria from the mother. Over time mitochondrial DNA has accumulated mutations which can be used to link women to lineages.

Careful studies utilizing mitochondrial DNA have shown that all women in sub-Saharan Africa belong to one of the first three branches of the mitochondrial DNA tree known as L1, L2 and L3. Lineages M and N, the two daughter lineages of L3, were the only ones to come out of Africa. Most men outside Africa carry a Y chromosome mutation called M168, suggesting that modern humans left Africa before this mutation occurred. Studies of current distributions of founder germ cell mutations also support the theory of the African origin of modern humans. Of interest, recent studies have shown that minimal interbreeding took place between *H. sapiens* and other hominins.

The period from 2.5 million years ago, when tool-making began, to ten thousand BCE is known as the Old Stone Age or Paleolithic. Its last phase, between ten and forty-five thousand years ago, is referred to as the Upper Paleolithic because of its characteristic stone tools. The subsequent period, called the Neolithic, was associated with the beginning of agriculture. This shift from the Paleolithic to the Neolithic is related to the decline in big game for hunter–gatherers, the end of the ice age, and the beginning of the relatively warm Holocene. During the ice ages, northern Europe and Asia were dry and cold, covered by glaciers, requiring significant adaptation or southward migration of resident populations. The sea level was about two hundred feet lower than now. This may have facilitated the migration of modern humans out of Africa by crossing the southern end of the Red Sea and eventually reaching Australia and the Americas.

During the Upper Paleolithic these modern humans developed a culture that included burial sites and cave art. The arts of the Cro-Magnons, as they are called, indicate a capacity for symbolism and the complex recreation of the outside world. Their burial sites indicate not only a belief in the afterlife, but also clear evidence of social stratification. As they expanded, they occupied the territory of the Neanderthals; there is some evidence of minimal interbreeding between the two species. Recent genetic studies suggest that 87% of Europeans descended from people that arrived before the end of the Pleistocene ice age, and only 13% descended from ancestors that migrated from the Near East about ten thousand years ago.

Modern humans developed a primitive culture in Asia that was similar to that in Europe. The origin of the mongoloid cranial structure in Asians is not clear. Of recent evolutionary development, this adaptation may have arisen as a result of cold weather. Asians domesticated the dog, which may have given them a survival advantage. They also colonized the Americas. Australasia already hosted humans as far back as forty thousand years ago. The ancestral population was well adapted to the climate. Australian aboriginal tribes remained isolated until the arrival of Europeans. They lived in small groups and never developed agriculture. The people from New Guinea are probably descendants of the same migration. They share many physical and cultural characteristics with the Australian aborigines.

There were several large migrations from Siberia to the Americas based on studies of Y chromosome variation between the people from both regions. The first one took place approximately fourteen thousand years ago and spread the Amerind family of languages from north to south. It explains the puzzling archeological findings of Monte Verde in Chile dated twelve and a half thousand years ago. A second wave came later and remained restricted to Na-Dene speaking tribes in North America. This may have been followed by a third migration of Eskimo-Aleut speaking groups.

The ability to communicate with language may have been a key factor that preceded and prompted the development of modern humans. There is no consensus about the ultimate origin of language. The descended larynx was formerly viewed as a structure unique to humans; however, it has been found in other species. It may have been adapted many years later to expand the variety of sounds produced by modern humans. The price we pay for this anatomical change is the possibility of choking to death, because we cannot simultaneous breathe and swallow.

The ability to communicate by means of a proto-language may have developed in archaic *H. sapiens* in order to negotiate social pressures and to transmit information among group members. The development of language may have been a gradual phenomenon or it could have been the consequence of a more abrupt change in human genes. Changes in the highly conserved FOXP2 gene, which codes for several transcription factors, may have been the last step in the development of modern speech. The products of this gene, discovered in 1998 in a family with severe speech impairment, are important for the coordination of facial musculature during speech.

The earliest sign of human cultural development was the increased complexity and sophistication of the stone tools and jewelry found in East Africa. There was a succession of cultures in Europe during the Upper Paleolithic. The cave paintings from Lascaux and other regions in Europe indicate a fairly developed form of artistic expression. Burial sites show evidence of belief in the afterlife.

At the end of the last glaciation, humans were driven to the agricultural revolution in an increasingly desiccated land depleted of the natural food supply. The transition from hunter-gatherers to farmers and herders was aided by a genetic mutation giving people the ability to produce lactase, and the possibility to digest milk throughout life. Humans started to learn to domesticate wild animals and plants and established townships. By selecting and growing a few species of edible plants and domesticating animals they were able to feed more people and abandon their nomadic existence. Early forms of religion included totems, ancestor worship and cults of the Sun, Moon and Earth. The new technologies that they developed brought about specialized labor and social changes. Competition for accumulated resources became a frequent cause of conflict between and within groups

Copper was the first metal to be discovered, about 4000 BCE, whereas iron smelting was not invented until approximately 1200 BCE. Copper smiths soon learned that "impurities" could be enhancements, and the Bronze Age was born. Agriculture developed in the fertile valleys of the Tigris, the Euphrates, the Nile, the Indus, the Yangtze and the Yellow Rivers. In Mesopotamia agriculture was already established by 6000 BCE; the invention of irrigation canals, the use of the sail and the wheel, and the establishment of the first city-states soon followed. The first codification of the law by Hammurabi dates back to 1750 BCE. The Egyptian civilization began about 3000 BCE when the country was united. In India, large cities like Mahuang-Dark and Harappa were well developed by 2500 BCE. The civilizations of China and Mesoamerica date back to 1500 hundred BCE.

The Sumerians in Mesopotamia were the first humans to develop a complete written language. Their cuneiform symbols were used to record commerce and business transactions. In Egypt hieroglyphics were used for written communication. Scientific studies concerning measurement of time and the movement of celestial constellations can be found in the Babylonian and Egyptian cultures. The latter used the sundial, telling the time of day by noting the angle of the sun's shadow. Megalithic monuments such as Stonehenge in Britain and the Pyramids in Egypt were probably designed for astronomical and religious observations.

The growth in population of agricultural societies led to the development of cities and states and the creation of a division of labor including workers, priests, soldiers and ruling classes. The possession of domestic animals, the yield of crops and the ability to travel long distances permitted an improvement in nutrition. The military use of horses and camels and the development of improved weapons led to wars of conquest and close proximity to animals may have been responsible for the spread of animal diseases into human populations. The availability of food surpluses and means of transportation propelled the development of stable, socially-complex, technologically-advanced societies.

When iron — a cheap, hard metal — became widely available, it revolutionized agriculture, industry and warfare. As the Bronze Age empires collapsed new geopolitical systems replaced the old empires. The great Iron Age civilizations included the Persian, Indian, Chinese, Greek and Roman empires.

Military victories against neighbors launched the golden age of Greek civilization. Many of their achievements during the fifth century BCE became the foundation of Western Civilization. There was an explosion in the arts, literature and the recording of history; the medical, scientific, and political innovations were unmatched in most of prior human history. Most importantly, Greece was the birthplace of participatory democracy, unfortunately limited to male citizens above age thirty, excluding slaves and women. The Greeks were polytheists and they worshipped gods that were anthropomorphic; they behaved much like men.

During the second century CE, the Roman Empire consolidated its territorial gains over the Greek and other cultures and became the most advanced civilization of its time. Subsequently, the Roman Empire became prey to invasions from the barbarians. By the end of the empire, Christianity had grown considerably and eventually became the official religion; the prevailing paganism had been almost completely eliminated. The final split of the Roman Empire between the Eastern and Western divisions took place in 395, after the death of Theodosius I. The western half was replaced by a patchwork of Germanic kingdoms but in the east Byzantine civilization grew and flourished.

After the fall of the Roman Empire, most of Europe plunged into the Middle Ages, a period characterized by feudalism, obscurantism and very few cultural advances. Islam, by contrast, achieved great territorial and cultural expansion during the seventh and eighth centuries and became a repository of many of the advances of classical civilization. Islam was spread by Arab armies through the Middle East, North Africa and Spain. In 1054 there was a schism between Eastern and Western Christianity, essentially a split between Catholicism and Orthodox Christianity. The next schism of Christianity took place during the sixteenth century at the time of the Protestant Reformation.

The fall of Constantinople to Ottoman Turks in 1453 caused human migrations and mingling of cultures that may have contributed, at least in part, to the upsurge in the arts and sciences and the rediscovery of the classical world that characterized the period we call the Renaissance. The Muslim kingdom of Granada was conquered by the Catholic monarchs Ferdinand and Isabella in 1492, the same year of the discovery of the Americas by Columbus, which marked the beginning of European voyages of discovery and colonialism. During the sixteenth and seventeenth centuries the Reformation and the Dutch and English revolutions made mercantile capitalism the dominant force in Europe.

At the end of the eighteenth century, ideas of social justice and political freedom began to circulate among European intellectuals, supported by Rousseau's writings including his *Social Contract*. Such Enlightenment ideas signaled the end of feudalism and monarchy and planted the seeds of the liberal democracies that flourished during the late nineteenth and the twentieth century. At the beginning of the nineteenth century the industrial revolution began in England; it was based on technological advances, such as the steam power, that permitted the mass production of consumer goods and the introduction of a new production system, the factory. This was the beginning of an era of technological advances such as the light bulb, the internal combustion engine, the telephone and the telegraph. An urban work force was created, as well as the capitalist economic system, and ushered in the beginning of European colonialism and dominance of the world. This new system in some ways conflicted with any idealistic view of a presumed social contract, prompting Marx and others to analyze its flaws and propose revolutionary ideas of their own. The industrial revolution was followed by the digital revolution which began in the early 1970s, after the introduction of computers, and the rapid growth of the Internet.

Chapter 6. Mind and Brain

Most living organisms have evolved a centralized nervous system which is responsible for sensing the outside world and responding appropriately. Mind is a term usually reserved for the complex behavioral and cognitive abilities associated with the function of the human brain. Nevertheless, some people believe that everything, including animals and inanimate objects, has both mental and physical properties (panpsychism). It is difficult to argue that a stone has a mind, however, most primitive organisms, even those without brains, have the capacity to detect changes in the environment, and to respond appropriately. For instance, some unicellular organisms exhibit molecular sensing mechanisms and are able to react to the environment.

We now know that the biological substrate of minds, the nerve cells or neurons, exists in many primitive cell types. Voltage-gated ion channels capable of creating electrical signals can be found in bacteria and archaea. Some species of *Paramecia* may exhibit action potentials under physical stimulation. Electrical excitability arose before neurons appeared.

Sponges and placozoans, the most primitive animals, lack neurons but contain the genetic blueprints for a set of proteins found in the synapses that connect nerve cells. Sponges have contractile cells (myocytes) that control the rate of nutrient-saturated water flowing through the animal. The larvae of some sponges express genes that control the formation of neurons from precursor cells; others may be able to generate action potentials. As it often happened during evolution, it appears that organisms co-opted pre-existing genes and functional capabilities to jerry-built the nervous system.

Cnidaria, a group of animals with radial symmetry that includes jellyfish, hydras, anemones and corals, are the most primitive animals in which we find neurons. Neurons have the capacity to transmit electrical messages utilizing long extensions that permit them to establish rapid communication with other cells. In primitive animals neurons formed a diffuse web, as evolution progressed these webs eventually coalesced to create more complex nodes and a centralized brain. Some of these organisms had cells capable of light perception, the creation of action potentials and the release neurotransmitters across synapses. For instance, Hydras have sensory neurons derived from the outer or ectodermal layer; these cells connect with motor neurons that innervate contractile cells. Their neurons form specialized neural nets that are close to the epidermis in order to permit a more rapid response to stimuli and support feeding and locomotor behaviors (DeSalle and Tattersall, 2012).

The evolution of bilateral symmetry in animals was associated with centralization into ganglia, nerve cords, nerves and plexuses, and regionalization and segmentation of the nervous system. In some invertebrates like the nematode *C. elegans* it is possible to identify ganglia or clumps of neurons and a tendency for these cells to concentrate in the head. These organisms have only about three hundred neurons, including sensory and motor neurons that have been mapped and studied in great detail. They permit the animal to carry out complex behaviors including finding food and mates and avoiding noxious stimuli.

More evolved invertebrates like insects, crustaceans and mollusks have more complex nervous systems. Among arthropods, fruit flies have brains containing about a hundred thousand neurons. Their brains exhibit mushroom bodies, which are special structures involved in odor perception and other functions, and optic lobes comprising almost half of the brain. Fruit flies exhibit complex aerial displays, mating dances and have been used to study reward behaviors. The brains of cephalopod mollusks have approximately one hundred and fifty million neurons. These animals are very intelligent and are known to navigate mazes, solve problems and learn complex behaviors.

The entire vertebrate nervous system is derived from the embryonic ectodermal layer that also generates the skin. The notochord induces differentiation of ectoderm into a neural tube. In the first stage of development the anterior half of the neural tube exhibits three sequential swellings the hindbrain, the midbrain and the forebrain. Of interest, the genes that control the formation of these structures is already present in primitive animals. The forebrain includes the cerebrum, the thalamus and the hypothalamus, the midbrain consists of the tectum and the tegmentum and the hindbrain gives rise to the pons, medulla and cerebellum.

There are several different types of brain structures among the vertebrates; the most notable anatomical change during the evolution of vertebrates was the progressive enlargement of the cerebral cortex. Fish have small brains with well-defined cerebellum, optic lobes and diencephalon. The brains of reptiles and birds contain a dorsal ventricular ridge, different from basal ganglia, which is felt to be part of the pallium, the equivalent of the cerebral cortex in mammals. Contrary to conventional wisdom, some birds like jays, parrots and corvids exhibit intelligent behaviors. The limbic system, clearly present in mammals, is the site of emotions, largely automatic actions and bodily changes designed to protect the organism.

The first hominins had a brain size of three hundred and twenty to three hundred and eighty cubic centimeters, about the size of an average chimpanzee. The genus Australopithecus lived in Africa between four and one million years ago. These hominins were bipedal, had dental features different from those of the great apes and possessed small brains. As mentioned above, the origin of the genus *Homo* remains elusive, early specimens appear about 2 million years ago. The "Turkana boy" had skeletal proportions similar to us and a brain size of nine hundred cubic centimeters. Modern humans, who appeared about two hundred thousand years ago, had a brain size of eleven hundred and fifty to thirteen hundred and twenty five cubic centimeters, close to present levels. Neanderthals also had a large brain and exhibited excellent motor skills, however, they were not as cognitively advanced as the "Cro-Magnons," the modern humans that lived in Europe thirty to forty thousand years ago.

It took several million years for the brain of our ancestors to expand to the size of modern men. During this period, the brain grew progressively; it now contains about one hundred billion neurons. Most of the growth took place in the cerebral cortex; the factors responsible for this rapid growth have not been clearly established. Some experts claim that social interaction, tool-making or climate changes influenced brain growth, and that this change was associated with the development of language, intelligence and consciousness. Another theory suggests that inactivation of the gene responsible for the chewing muscles in most primates may have removed a constraint on brain growth. Finally, the selection of a gene known to be a specific regulator of brain size, such as the microcephalin gene, could have favored brain growth. Together with cultural change and transmission, the increase in brain size and connectivity may have been responsible for our spectacular success as a species.

Evolutionary theory suggests that complex but anatomically different brains have evolved multiple times independently during evolution. The concept of a linear progression in the complexity of the brain from invertebrates to vertebrates and ending in the human brain is no longer accepted by many experts. The current notion is that the human nervous system is not hierarchical, but instead, it is a distributed and interactive network.

The topographic model of the human nervous system can be supplemented with a model based on function (Swanson, 2012). The nervous system influences behavior through the motor system, which in turn receives inputs from the cognitive, sensory and behavioral state systems. The motor system has three components: the somatic motor system that controls voluntary muscles, the autonomic nervous system with its two branches, the sympathetic and parasympathetic nervous systems that innervate smooth muscle and glands and the neuroendocrine system that controls the hormonal output of the hypothalamus. The cerebellum is a part of the motor system involved in coordination and motor learning.

The cerebral cortex is the anatomical site of the cognitive system. The basal ganglia, the striatum and pallidum, also part of the cerebral hemispheres, modulate the effect of the cognitive system on the motor system. The cornerstone of our complex cognitive processes is located in the pre-frontal cerebral cortex. This region allows us to, plan, decide, make predictions, draw inferences and hold images in the working memory. Here information from our senses and other systems are combined with memory in order to establish high-order associations. Language provides a unique way to categorize the world and permit us to communicate with each other. Consciousness is a state of mind in which we are aware of ourselves and our surroundings.

The sensory system for the primary modality classes (somatosensory, vision, audition, olfaction and taste) is arranged in parallel fashion. In addition, there are other areas that provide sensory input for the regulation of fluid balance, eating and drinking. The subfornical organ has angiotensin II receptors and its output helps regulate blood pressure, modulates the release of vasopressin and connects with hypothalamic osmoreceptors that regulate thirst and water balance. There are neuron populations in the arcuate nucleus of the hypothalamus and other areas that regulate eating behavior and energy balance by responding to circulating hormones like insulin, leptin and ghrelin.

Somatosensory pathways convey information about touch, pressure, temperature, pain, vibration and position. The information is relayed to nuclei in the thalamus and integrated in the sensory cortex in the parietal lobes to provide conscious awareness of sensation. Information is also transmitted to other cortical neurons to adjust fine movements and body posture. Some of the ascending sensory fibers enter the midbrain and project to the amygdala and limbic system thus providing a pathway for the emotional response to sensory stimuli. In the spinal cord, painful stimuli elicit firing of lower motor neurons resulting in reflex withdrawal of the affected part.

Capsaicin is a chemical that is responsible for the effects of hot peppers. Its receptor is a member of a family of proteins called "transient receptor potentials" or TRP channels. These proteins are located in the outer membrane of nerve cells and when activated they allow a flood of calcium to enter the cell. TRP channels have recently been found to enable vision, taste, olfaction, pain and temperature.

The visual system provides one of the most important sources of information about the environment. It starts with the capture of images focused by the cornea and lens in a light-sensitive membrane, the retina, in the back of the eye. The retina is an extension of the brain that transduces light energy into neuronal signals. Ganglion cell axons exit the eye at the optic discs and travel through the optic nerves. Fibers from the nasal half of the retina cross to the other side in the optic chiasm and continue as the optic tract to reach the lateral geniculate nuclei of the thalamus. The visual pathway ends in the visual cortex located in the occipital lobes. The visual cortex, in turn, originates fibers that travel in the opposite direction and reach the lateral geniculate nuclei.

The posterior retina contains two classes of specialized photoreceptor cells, rod and cones, which transduce photons into electrical signals. The rods are sensitive to low levels of light and are more numerous in the peripheral zone of the retina. Color vision depends on cone cells. Color is not a property of light, or the objects that reflect light; it is an experience that is somehow created in the brain. Birds, turtles and many fish have four types of cones. Humans and some primates have three types and most mammals have two types of cones. Each cone has a pigment that is sensitive to a specific wave length of light. Of interest, for the brain to detect color it has to compare the responses of two or more types of cones. It has been suggested that early mammals lost their cones by being nocturnal and using primarily their rods to see around in the darkness. Humans and some primates may have gained back one type of cone by a random mutation that took place in a common ancestor.

Sound waves set the tympanic membrane in motion. Movement of the small chain of little bones in the ear results in changes in pressure in the fluid-filled inner ear. Movement of the hair cells in the cochlea elicits changes in ion fluxes that result in an action potential in the eighth nerve. These fibers are relayed through the midbrain to the auditory cortex in the superior temporal gyrus.

The vestibular system end-organs that keep us in balance are located in the semicircular canals and the otolithic apparatus (utricle and saccule). The canals transduce angular acceleration whereas the otoliths transduce linear acceleration and static gravitational forces. The neural outputs from the end organs travel to the vestibular nuclei in the brain stem via the eighth cranial nerve. The projections from the vestibular nuclei travel to other cranial nerve nuclei, the cerebral cortex and the cerebellum. The cerebellum is primarily concerned with balance, muscle tone and coordination of motor activity.

The sense of smell determines the flavor and palatability of food and helps monitor potential mates, predators and inhaled toxic chemicals. In humans a large family of genes is collectively used to make receptors to detect a great variety of environmental odorants. The olfactory neuroepithelium is located in the superior part of the nasal cavities. Electrical signals are generated when odorants bind the olfactory receptors expressed in the dendrites of olfactory receptor neurons. Their axons terminate in the olfactory bulb.

The taste receptor cells are located in the taste buds. There are five basic tastes: sweet, sour, bitter, salty and umami, a savory flavor exemplified by the amino acid glutamate. Salty and sour perceptions are mediated by ion channels and the rest by G-protein coupled receptors. The sense of taste travels in the ninth and tenth cranial nerves to the brain stem nucleus of the tractus solitarius. From there the fibers reach the cortex by different pathways.

The most important behavioral state system is the circadian rhythm. Throughout evolution living organisms have adjusted their physiology to the day–night cycle caused by the Earth's rotation. A circadian molecular clock is also present in all cells in the body. In mammals the control circadian pacemaker is found in the suprachiasmatic nucleus of the hypothalamus. It maintains the circadian cycle even in animals placed in constant light or darkness. Wakefulness depends on a network of cell groups in the brain stem, the hypothalamus and the basal forebrain that activate the thalamus and the cerebral cortex. These neurons are inhibited during sleep by several cell groups. Mutual inhibition between arousal and sleep-producing circuits creates a switch that determines the level of arousal. During sleep the cognitive system may be active as dreaming occurs, but the sensory and somatic motor systems are inhibited.

Like most complex animals humans exhibit a reproductive cycle, which tends to maximize the production and survival of offspring. Women exhibit approximately a lunar cycle. The reproductive cycle is triggered by the release of gonadotropin-releasing factor by neurons in the hypothalamus. This factor stimulates the production of gonadotropic hormone by the pituitary gland which in turn induces ovulation and the secretion of sex hormones.

Our basic survival-oriented functions are automatically regulated by the brain. The brain centers that control respiration, circulation and the acidity of body fluids are located in the medulla. The functionally-related hypothalamus regulates the endocrine system, circadian rhythms, arousal, energy balance and the autonomic nervous system. The nervous system plays an important role in the origination of homeostatic responses to internal states, such as hunger, thirst, pain and fatigue, or to external stimuli that elicit fear, anger, joy or sadness. Emotions provide the brain with a natural mechanism to evaluate the environment within and around the organism and to respond accordingly. Feelings are constituted by the perception of a certain body state and are the substrate for more advanced cognitive mechanisms.

Consciousness and the Body–Mind Problem

Monism is the metaphysical concept about the nature of reality that claims that there is only one substance. It includes physicalism or materialism, the view that there are only physical entities, and idealism which holds that reality is not physical. Dualism postulates the existence of both physical and non-physical entities. Substance dualism affirms that we have both a non-physical mind and a body and that they both exist apart from each other. Since they have two different set of properties they represent two kinds of substance. Because of the problems associated with both physicalism and dualism some philosophers have embraced Russellian monism, the view that the universe contains some special properties that underlie, and are more basic than, both mental and physical properties.

The most important problem of the philosophy of mind is the body-mind problem, the relationship between the mental and physical properties. There are two aspects of this problem that defy a purely scientific or physicalist explanation. One is consciousness, the felt subjective quality of experience and the other is intentionality, the capacity of the mind to be directed at something. Many experts in the field assert that our mental life is thoroughly embodied. Nonetheless, there is no clear definition of what "embodied cognition" represents.

Many religions embrace substance dualism and claim that the mind is a manifestation of a non-physical soul; others hold that they are separate but related entities. Although the soul means different thing to different people, most people consider it the eternal and deepest level of being, the site of our experiences and higher functions. A problem with dualism is causal inertia; the mind may be unable to affect the world outside it. There are no rigorous scientific observations that demonstrate that we possess an immaterial soul or a type of matter that defies known physical laws.

In the past many people claimed, without evidence, that living organisms exhibit other non-physical properties, such as a vital force, that is absent in inanimate objects. Extrasensory perception, the ability of minds to communicate with one another by unknown mechanisms, and psychokinesis, the ability of thoughts to influence physical objects, have not been confirmed by experimental studies. The initial claims that prayer can affect health outcomes have not been corroborated by larger and carefully performed studies carried out in patients with coronary heart disease.

The ancient Greeks viewed the body as containing channels for the soul or spirit, the force that gave a body life and thought. Aristotle placed the soul in the heart. Following Galen, the church fathers located the soul in the empty spaces in the head, disregarding the teaching of Epicurus who challenged the notion of the soul. Descartes claimed that he didn't have any doubt he had a mind, but could doubt that he had a body and therefore concluded that mind is distinct from the body, we have a non-physical mind or soul, the res cogitans, and a physical body.

Descartes' argument, though, is guilty of the Intentional Fallacy, because it treats doubt as though it were a property of the thing doubted. Descartes' second argument for dualism affirms that mind is indivisible but the body is divisible, thus mind is not identical to the body. This argument has also been criticized because it begs the question; it assumes the truth of the conclusion in the premises. It is unclear where and how the soul could interact with physical and non-physical entities. If the mind cannot occupy space, there can be no place in the brain for the interaction to take place.

Descartes postulated that this interaction took place in the pineal gland but he didn't explain how the interaction occurred. Proponents of parallelism held that that the mind and the body operate in parallel and that this relationship was arranged by God. A more recent response to interactionism is a view called epiphenomenalism. In this view, the brain has the power to produce non-physical states but mental states cannot cause brain states. Mental states are a special kind of causal by-product of events in the brain.

In the seventeenth century Willis created the first detailed maps of the brain and traced the nervous system throughout the body. Current studies using modern imaging techniques have shown that recognizing our bodies and our self-concept are accompanied by activation of two different parts of the brain. These findings may be the reason why we experience mind and body separated from one another. Substance dualism has been abandoned by most contemporary scientists and philosophers who have turned toward some form of physicalism. The reason why substance dualism is still held by a large percent of the population is the influence of religion.

Physicalism is the ontological theory that everything is physical. An important concept in physicalism is that of supervenience, the idea that higher levels of existence are dependent on lower levels, such that there can only be a change in the higher level if there is also a change in the lower level. There are three types of physicalism. Eliminative materialists attempts to do away with efforts to identify mental states with physical states. Instead they adopt the view that there are no such things as mental states. Reductive physicalism allows that some mental features exist but insist that they represent nothing but physical features. Non-reductive physicalism holds that although mental features are different from physical properties, the former are determined or "metaphysically supervene" upon the latter.

One of the first reductive physicalist theories was behaviorism, it arose in the wake of logical positivism, a view designed to avoid metaphysical questions that have no apparent solutions. Behaviorists captured some important aspects of the relationship between mental states and behavior but they were mistaken in their attempt to identify mental states solely with behavioral dispositions. Identity theorists admit that we have thoughts and sensations and claim that such states are identical to more fundamental neurophysiologic phenomena. Many philosophers think that the principle of multiple realization; the observation that the same type or kind of mental state can exist in a variety of physical systems, is a genuine obstacle to conceiving of the identity between the mental and the physical as an identity between types. Thus it has been suggested that we should adopt a form of physicalism that holds that each token of a psychological state is in fact a token of a neurophysiologic state.

Functionalism claims that functional definitions could succeed where the identity theory has failed. Thus to be in pain is to be in a state that is caused by bodily injury and causes pain behaviors. To define pain in terms of its causes and effects is to define it in terms of its intrinsic properties. Functionalism claims that mental states are multiple realizable, it supports the views of artificial intelligence. Nevertheless, its critics have raised arguments that drive a wedge between functional organization and the qualitative character or qualia of mental states.

It has been suggested that we should adopt a form of physicalism that is non-reductive. Although the brain is all there is to the mind, according to emergentism the description of mental events cannot be reduced to the lower level explanations of the physical sciences. Emergent physicalism has been introduced to explain how new forms of organization may emerge that are not explainable in terms of lower levels laws and processes.

Davidson's anomalous monism claims our having thoughts or sensations are events that have an alternative physical description. This means that mental events are not a different class of things, they are simply physical events described using the vocabulary of psychology. He denies that there are strict deterministic laws connecting mental events with physical events. This theory has been criticized because it appears to lead to a form of epiphenomenalism stripping us of any genuine rational agency.

It is difficult for physicalism to reconcile their notions of the physical with the experiential, how electrochemical processes in the brain result in subjective or first person experiences. Qualia, often referred to as phenomenal properties, are the subjective aspects of experience, the way sensations feel to those who have them. Postulating that qualia are poorly understood non-physical and unknown aspects of the natural world gets us no closer to understanding them. Current theories don't explain how a physical structure like the brain tissue creates the alluring images of a seascape or the romantic mood of a melody. It is also unclear if the soul exists and how the soul can influence physical events in the outside world.

Chapter 7. Homeostatic Regulation

Science has been able to explain most of the basic systems that are responsible for keeping us alive. Some of these systems help us to solve automatically the basic problems of life such as finding nutrients, maintaining an internal structure and chemical balance, fending off agents that cause disease or physical injury and reproducing. The respiratory system is designed to retain oxygen from the inspired air and the digestive system digests and absorbs ingested nutrients. The acquisition and storage of proteins, lipids and carbohydrates is necessary to maintain structure, and to provide energy for bodily functions. Waste products are eliminated via the lungs, kidneys and the gastrointestinal tract. Defense mechanisms include emotions, basic reflexes, like the startle reflex, the stress response, the behaviors triggered by sensory experiences and the function of the immune system. The reproductive system assures the continuation of the species.

Communication between cells in the body is mediated mainly by the nervous system with the contribution of the endocrine, circulatory and immune systems. These systems constitute an interlocking network. Not only hormones circulate in the blood but also neural impulses have major effects on the release of certain hormones. The immune system is under nervous and endocrine control and cytokines produced by lymphocytes may affect endocrine and brain function.

The hypothalamus is a brain structure that is ideally located to maintain the constancy of the internal environment and many of the basic mechanisms of life regulation. It is the part of the brain where the activity of endocrine glands, the immune system and the autonomic nervous system is integrated with input from higher centers and where the regulation of emotions and behavior takes place.

The anterior pituitary gland, appropriately titled the master gland, produces six major hormones and stores an additional two hormones. It is connected to the hypothalamus by a stalk at the base of the brain. It acts as an intermediary organ, receiving hormonal inputs from the hypothalamus and connecting with the body through hormonal products that influence other endocrine glands. In humans, anterior pituitary cells produce the following hormones: adrenocorticotropin hormone (ACTH), thyrotropin hormone (TSH), growth hormone (GH), prolactin, luteinizing hormone (LH) and follicle-stimulating hormone (FSH). ACTH controls the function of the adrenal cortex, TSH regulates thyroid function, GH regulates growth and influences intermediary metabolism, prolactin is necessary for lactation, and LH and FSH regulate gonadal function. Two additional hormones, oxytocin and vasopressin are produced by neurons in the hypothalamus and stored in the posterior lobe of the pituitary. The former controls water conservation by the kidneys and the latter is necessary for milk let-down during lactation.

The autonomic nervous system innervates vascular and visceral smooth muscle, exocrine and endocrine glands and parenchymal cells in various organ systems. It responds automatically to perturbations that threaten the constancy of the internal environment such as changes in blood flow, tissue perfusion, blood pressure, volume and composition of the extracellular fluid, energy balance and smooth muscle tone. Autonomic neurons located in ganglia outside the central nervous system give rise to post ganglionic fibers that innervate body tissues. This activity is regulated by central neurons responsive to diverse afferent impulses. After central integration autonomic outflow is adjusted accordingly.

The sympathetic and the parasympathetic are the two main branches of the autonomic nervous system. Sympathetic outflow is initiated from the reticular formation of the pons and medulla and from centers in the hypothalamus that synapse in the spinal cord with preganglionic sympathetic neurons. The preganglionic fibers of the sympathetic nervous system exit the spinal cord between the first thoracic and the second lumbar segments. The preganglionic fibers of the parasympathetic leave the central nervous system in the third, seventh, ninth and tenth cranial nerves and in the second and third sacral segments. Acetylcholine is the preganglionic neurotransmitter in both divisions as well as the postganglionic neurotransmitter of the parasympathetic nervous system. Norepinephrine is the neurotransmitter for most of the postganglionic sympathetic nervous system. Response to stimuli by these two systems is frequently antagonistic, thereby providing a more precise control of autonomic responses.

The immune system protects the host against injury caused by foreign agents. We have an innate system of natural immunity and an adaptive system that is acquired. The former is the first line of defense against invasion by infectious agents. Pattern recognition receptors such as Toll-like receptors detect highly conserved microbial-specific structures. Natural immunity includes acute phase proteins, the alternative complement system, and phagocytic, dendritic and natural killer cells. Some of these are also involved in acquired immunity; there is substantial overlap between the two systems.

The red blood cells in the circulation transport oxygen to the metabolizing tissues and the white blood cells are involved in the defense against pathogens. The adaptive immunity consists of antibody-mediated and cell-mediated immunity. The former depends on the production of immunoglobulins by B-lymphocytes. These proteins bind to the invading agent to inactivate it or to make it susceptible to phagocytic cells or attack by the complement system. The cell-mediated system consists of T-lymphocytes that express surface antibody-like receptors that recognize the invading agent in the surface of antigen-presenting cells. The latter exhibit class I or II main histocompatibility complex (MHC) molecules that bind fragments of the invading antigen for presentation to the T-cells. Cytotoxic T-cells exhibit class I MHC and helper T-cells MHC class II molecules.

Immune cells manifest antibody specificity before exposure to immunizing agents; antigens serve to stimulate their proliferation (clonal selection theory). The body has the capacity to synthesize almost a limitless number of antibodies in response to foreign antigens. The immune system is under neural and hormonal control; it is also self-regulated to avoid auto-immunity. Lymphoid cells have receptors for hormones and are influenced by the autonomic nervous system. Products of the lymphoid cells such as cytokines are known to activate the pituitary-adrenal axis and glucocorticoids, in turn, are capable of influencing lymphocyte function.

Wakefulness and Sleep

Wakefulness depends on the activation of a variety of brainstem nuclei. These nuclei include cholinergic nuclei of the pons-midbrain junction that project to thalamo-cortical neurons, noradrenergic neurons of the locus coeruleus and serotonergic neurons of the raphe nuclei. The system is activated during wakefulness and rapid eye movement (REM) sleep. The firing of these neurons is periodically inhibited by neurons from the ventrolateral preoptic nucleus of the hypothalamus. Inactivation of this system results in partial or complete loss of wakefulness. Wakefulness is one of the components of consciousness. It is necessary for attention, the faculty that amplifies relevant information and inhibits distracting inputs, and is closely linked to the reward pathways.

Sleep has been observed to take place in many animals ranging from insects to humans. Its function has not been elucidated. Sleep occurs with circadian (about one day) periodicity, even in the absence of external clues. REM sleep in man is associated with dreaming, emotion, visual hallucinations and lack of volitional control. Non-REM sleep is characterized by decreased muscle tone, heart rate, breathing and metabolism. To synchronize sleep with the day–night cycle specialized cells in the retina containing the pigment melanopsin project to the suprachiasmatic nucleus (SCN) of the hypothalamus. This site connects with pathways involved in other functions that are also synchronized with the day–night cycle such as temperature, hormone secretion and blood pressure changes. Activation of the SCN evokes responses in sympathetic neurons in the spinal cord and superior cervical ganglia that project to the pineal gland. This gland synthesizes the sleep-promoting hormone, melatonin. Of interest, most body tissues have an intricate molecular mechanism to mark the circadian cycle. The SCN is just the master regulator in the brain acting from light inputs from the outside world.

Temperature Regulation

Like all mammals, we exhibit the capacity to maintain a constant body temperature despite a wide range of ambient temperature. The afferent loop of this system consists of central and peripheral temperature receptors, the integration takes place in the hypothalamus, and the efferent loop resides in the autonomic, endocrine and musculoskeletal systems. Heat conservation or elimination is aided by changes in the caliber of peripheral vasculature. Thus vasodilatation helps to remove heat from the body and vasoconstriction has the opposite effect. Increased heat generation above the basal level is mediated by increased production of ATP or reduction in the efficiency of ATP synthesis. The set point for the system is around 37 degrees Celsius. It can be increased by endogenous pyrogens or decreased by drugs such as aspirin.

The Stress Response

Competition for mates and resources is an important source of stress for all living organisms. Establishing and maintaining dominance in a group activates the stress response for both the dominant and the subordinate individual. Stress elicits a spectrum of physiologic changes designed to prepare the organism to "fight or flight." Prominent among these changes are the release of glucocorticoids by the adrenal gland and activation of the sympathetic nervous system. Glucocorticoid secretion is accomplished by the hypothalamic-pituitary-adrenal (HPA) axis. Upon stress, the neurons of the paraventricular nucleus (PVN) of the hypothalamus secrete corticotropin-releasing hormone and vasopressin. The release of adrenocorticotrophic hormone drives the synthesis of glucocorticoids by the adrenal medulla.

Excitation of the HPA axis is driven by several central stress circuits. Among these are catecholamine-producing pathways which project directly to the PVN nucleus. The bed nucleus of the stria terminalis may also convey stimulation to the HPA axis. This limbic forebrain structure connects regions such as the amygdala and hippocampus to hypothalamic and brainstem regions controlling vital homeostatic functions. Limiting the stress reaction is the negative feedback exerted by glucocorticoids on the PVN. The PVN neurons also receive direct inhibitory input from the limbic system, the hypothalamus and the prefrontal cortex. Chronic elevation of stress hormones stress may cause adverse health effects including the development of cardiovascular and autoimmune diseases.

Energy Balance

The maintenance of energy balance is another vital function orchestrated in the hypothalamus. Energy intake (feeding) must equal energy expenditure (basal metabolism, physical activity and thermogenesis) for body weight to remain stable. The long-term control center for energy balance in humans is located in the hypothalamus. It receives numerous afferent signals such as leptin (adipose tissue), ghrelin (stomach) and insulin (pancreas) and originates changes in food intake and energy expenditure. Whereas leptin inhibits food intake and increases energy expenditure, ghrelin induces a positive energy balance by promoting food intake and decreasing energy expenditure.

The arcuate nucleus of the hypothalamus contains two populations of leptin-sensitive neurons. Some neurons promote weight loss and are activated by leptin. They release the anorexigenic peptide alpha-melanocyte stimulating hormone (alpha-MSH) that binds to special receptors and leads to suppressed food intake. At the same time, gherlin stimulates and leptin suppresses other neurons that release orexigenic (appetite-inducing) peptides. Activation of the arcuate nucleus also promotes energy expenditure by stimulating ACTH and TSH release and activating brain stem nuclei associated with the sympathetic nervous system. It may also influence other nuclei in the lateral hypothalamus to produce orexigenic stimuli such as orexins. Orexins regulate sleep and wakefulness by excitatory projections to arousal centers (low levels are found in the sleep disorder narcolepsy) and interact with the reward system.

Short-term regulation of energy balance takes place during the cephalic, gastric and substrate phases of digestion. Gastric distension inhibits feeding via the vagus nerve and the nucleus of the tractus solitarius in the medulla. Hormones released by the gut have hunger and satiety-stimulating actions. Fatty foods release cholecystokinin from the intestine causing vagal inhibition of food intake. Insulin and glucose increase substantially during the substrate phase. Insulin induces satiety acting on the hypothalamus.

Digestion

Digestion involves the breakdown or hydrolysis of nutrients to small molecules to prepare ingested substances for absorption or transport across the intestinal cell. The digestive process starts in the mouth where food is triturated and exposed to saliva. The motor activity of the esophagus propels food to the stomach. Most of the digestive process takes place in the stomach by the action of acid and pepsin and is continued in the upper small intestine primarily by pancreatic enzymes such as lipase, amylase and trypsin. As a result of these enzymes ingested carbohydrates are broken down to monosaccharides and disaccharides, proteins to peptides and amino acids, and fats to monoglycerides and fatty acids. It is in this form that most nutrients are absorbed in the small intestine. Water and sodium reabsorption in the intestine and the colon results in a stool volume of less than three hundred grams per day.

The mucosal surface of the gastrointestinal system is composed of a dynamical population of epithelial cells that have a high capacity for absorption and secretion. These cells facilitate digestion and nutrient uptake and at the same time maintain a barrier between the host and potential pathogens in the lumen. The intestinal tract contains a large population of immune cells; the predominance of suppressor lymphocytes damps the potential stimulatory effects of the wide variety of antigens present in the lumen. Other immune cells are ready to respond to foreign antigens when surface defenses have been breached. The mucosa of the gastrointestinal tract exhibits a very rapid turnover of epithelial cells which permits a rapid restitution of the resident cell population after an acute insult.

Fluid and Electrolyte Balance

Water is one of the most important requirements for life because most chemical reactions in living organisms take place in aqueous solution. The transition from an aqueous to an arid environment required the development of mechanisms designed to maintain water balance. The mechanisms that maintain water balance within a narrow range, in spite of large variations in the rate of intake and losses, are thirst and the secretion of vasopressin, the hypothalamic hormone that promotes renal water conservation.

The major physiologic stimulus for thirst and vasopressin secretion is the increase in plasma tonicity associated with dehydration. Volume contraction, nausea and a number of pharmacologic agents are non-osmotic factors that stimulate vasopressin secretion. The sensors for thirst and vasopressin release are in the hypothalamus, in close proximity to each other. Contraction of extracellular fluid volume also promotes thirst by baroreceptor-induced alterations of afferent parasympathetic tone and increases in angiotensin II levels. The latter activates thirst via its interaction with paraventricular structures such as the subfornical organ.

The evolution of the kidneys is closely tied to the evolution of mechanisms for the regulation of the electrolyte and water content of the body under different environmental conditions. We are mostly water; it constitutes about two thirds of the body weight. Total body water is distributed between the intracellular (three fourths) and the extracellular (one fourth) compartments. These compartments are separated by the cell membranes.

One fundamental property of living cells is the existence of pumps that extrude sodium ions from the cell interior across the cell membrane in exchange for potassium ions. Thus sodium and associated anions are the main osmotic particles in the extracellular fluid, whereas potassium and organic anions are the predominant ions inside the cells. Because osmotic equilibrium exists between body fluid compartments, the size of each compartment is determined by the number of osmotic particles present in it.

The extracellular fluid is subdivided into the interstitial fluid and the intravascular compartment; they are separated by the capillary wall. The hydrostatic pressure generated by the heart promotes the movement of fluid from plasma to interstitial fluid; this force is opposed by the osmotic pressure generated by proteins inside the capillaries. The volume of the extracellular fluid compartment depends on the total body sodium content. Volume sensors located in different organs monitor the volume of this compartment and instruct the kidneys to retain or to excrete sodium. Because sodium balance is regulated by the kidneys, these organs play a central role in the maintenance of extracellular fluid volume and therefore, the integrity of the circulation and blood pressure.

Potassium balance is also regulated by the kidneys. In addition to regulating intracellular volume, potassium, or more accurately, the gradient between intracellular and extracellular potassium concentration, determines the electrical potential difference across cell membranes that underlies the normal excitability and contractility of muscle and the electrical properties of neurons.

To maintain acid–base balance the lungs have to eliminate the carbon dioxide produced during cellular respiration and the kidneys must excrete the amounts of fixed acid produced as the result of the metabolism of ingested protein. The latter is accomplished by hydrogen ion pumps located in the renal tubular epithelium. As mentioned above, water balance is maintained by the thirst mechanism and by the secretion of vasopressin, a hypothalamic hormone that promotes water conservation by the kidneys.

The kidneys are two bean-shaped organs, each the size of a fist, located in the back on each side of the spine. Each kidney contains about one million microscopic units called nephrons. A nephron consists of a filtering bed composed of a capillary tuft, or glomerulus, which drains into a long and elaborate series of tubules that end in the collecting duct. The tubular system includes the proximal tubule, the loop of Henle and the distal tubule. The collecting duct, of different embryological origin, drains into the renal pelvis which then connects to the bladder via the ureters.

The kidneys receive about one fourth of the cardiac output and filter approximately one hundred and twenty milliliters of water and dissolved small molecules per minute. Large molecules such as proteins and blood cells are not filtered and remain in the bloodstream. After leaving the glomerulus, the blood passes into a second set of capillaries surrounding the tubule. Valuable substances in the tubular fluid, such as glucose and amino acids, are reabsorbed; other substances are secreted from the blood into the tubular fluid. Waste substances present in the original filtrate and those secreted into the tubule are excreted in the urine. When the kidneys fail, toxic waste products accumulate in the blood, renal hormones are not produced, and fluid, electrolyte and acid–base balance is impaired.

The excretory systems of invertebrates differ greatly in their structure and complexity from that of humans. Flat worms have a branched system of tubules throughout the body. Fluid in tissues passes through a bulbous cell into an excretory tubule that connects with the outside. Round worms have tubules that connect the body cavity with the outside and insects have Malpighian tubules that open into the digestive tract.

Ancestral vertebrates that lived in the ocean drank sea water, bathed their tissues in it, and then expelled the residue through a simple conduit. As they migrated to fresh water they retained an internal environment of high salinity. The consequent flow of water into these animals posed the risk of overhydration and created the need to have an organ able to excrete large amounts of water. This organ was the glomerular kidney, each glomerulus consisting of a network of capillaries ending in tubules permitting the filtered fluid to drain to the exterior.

Some experts claim that glomerular kidneys already existed in ancestral marine vertebrates, whereas others claim that migration to fresh water created the evolutionary pressure to select an organ capable of water excretion. Regardless of their evolutionary history, glomeruli filter large amounts of essential small solutes that have to be reabsorbed by the proximal segment of the renal tubule along with osmotically-obligated water. Some water and solutes can also be reabsorbed in the distal segments: the loop of Henle, the distal tubule and the collecting duct.

Vertebrates that evolved in the oceans had to develop a method to minimize water losses due to the osmotic forces generated by the high salinity of the surrounding medium. Elasmobranchs (rays and sharks) adapted to the marine environment by raising the urea concentration in body fluids to match the high salt content of the oceans. Bony fish adjusted by excreting mineral salts through kidney tubules and gills, minimizing the flow of blood through the kidneys and making ammonia the main nitrogenous waste product. Fresh water fish avoided drinking water and their skin became fairly impermeable to water, any retained water was eliminated by the large filtration capacity of their kidneys. Nitrogen waste was excreted primarily as ammonia by diffusion through the gills.

When vertebrates migrated to land they faced the risk of desiccation. Reptiles solved the problem by reducing the glomerular size and function and by excreting uric acid as the end product of nitrogen metabolism. Uric acid is excreted as a semisolid paste with minimal amounts of water. Mammals maintained glomerular filtration and excreted urea as the main nitrogenous waste product. They avoided desiccation by drinking in response to thirst and by reabsorbing water through the action of vasopressin on the collecting duct along the osmotic gradients created by the renal countercurrent system. Birds adopted both systems, they excrete uric acid and some of their tubules have long loops of Henle.

Cardiorespiratory Function

The lungs are responsible for the uptake of oxygen from the inspired air and the removal of the by-product of metabolism, carbon dioxide. The lungs perform these functions by placing inspired air close to the pulmonary capillary bed to permit gas exchange by simple diffusion. Respiratory muscles are under automatic and voluntary control. The rhythm of spontaneous breathing originates in groups of interconnected neurons in the medulla. Their integrated output is an efferent signal that travels via the phrenic and spinal nerves and results in contraction or relaxation of the respiratory muscles. Under normal conditions, the hydrogen ion concentration in the region of the central chemoreceptors in the brain stem determines the drive to breathe. Peripheral chemoreceptors increase the respiratory drive only during moderate to severe hypoxemia.

The maintenance of the normal hydrogen ion concentration in body fluids is essential for cellular function. To maintain acid-base balance, the lungs have to eliminate the load of carbon dioxide generated in peripheral tissues and the kidneys must excrete the amount of nonvolatile acid produced during metabolism. In addition, the kidneys prevent the loss of bicarbonate (buffer) in the urine by reclaiming all the amounts filtered. The last two processes are accomplished by hydrogen ion secretion by pumps located in the renal tubular epithelium.

Given the vital nature of the cardiovascular system in maintaining blood flow to vital organs it is not surprising that we have evolved multiple cardiovascular regulatory mechanisms. Adjustments can be made changing the output of the pump (heart) or altering the resistance (diameter) of the blood vessels. Cardiac output is the product of the heart rate times the volume of blood pumped with each heartbeat. The latter depends on the contractile state of the heart, the impedance against which the heart pumps (afterload) and the amount of filling of the ventricle before contraction (preload). The caliber of the blood vessels is controlled by vasoactive compounds produced by metabolically active tissues, by autoregulation, by vasomotor nerves and by vasoactive substances produced by the endothelium and other tissues. Vasoactive agents include nitric oxide, angiotensin II, catecholamines and prostaglandins.

The main control of vasomotor discharge is feedback regulation by baroreceptors in the walls of the heart and central blood vessels. The signal is carried by the ninth and tenth cranial nerves to the nucleus of the tractus solitarius. Secondary order neurons pass the signal to the ventrolateral (VLM) medulla. From the rostral VLM, axons descend to sympathetic neurons in the spinal cord and from there to blood vessels, to the heart and to the adrenal medulla. Increased afferent baroreceptor activity inhibits the sympathetic outflow. The associated vagal discharge results in decreases in heart rate and cardiac output. In contrast, a decrease in baroreceptor afferent discharge stimulates sympathetic activity and inhibits vagal discharge.

Sex and Reproduction

The drive to reproduce is a basic behavior designed to assure the survival of the species. It is reinforced by pleasurable sensations associated with activation of the reward system. Sexual reproduction has been favored by evolution because it results in better recombination of genetic material and it enhances the ability to eliminate damaged DNA. The sex genotype is determined at the time of conception. Females carry two X chromosomes and males one X and one Y chromosome. Early in development one of the X chromosomes in females is silenced in order to maintain the same dosage of genes as in the males.

Gonadal sex is determined after 40 days of gestation by the presence or absence of a single gene, the SRY gene in the Y chromosome. This gene codes for a protein that promotes testicular development. In the absence of the testis, the organism develops along female lines.

Sex phenotype in humans, consisting of internal and external genitalia and secondary sexual characteristics, depend on the production of sex hormones (estrogens and testosterone) by the ovary and the testes. Gender is more complex than genetic or gonadal sex determination. It depends in great part on the individual's subjective perception of his/her sexual identity. Sexually-related differences in phenotypic expression of the genotype are called sexual dimorphisms. They are due in great part to the early influence of hormones (especially estrogens) on maturing brain circuits.

In the adult, the level of sex hormones is regulated by the hypothalamus in men and the hypothalamus and corpus luteum in women. Gonadotropin-releasing hormone produced in the hypothalamus stimulates the release of luteinizing hormone (LH) and follicle-stimulating hormone (FSH). LH stimulates the testicular production of testosterone and ovarian steroidogenesis. FSH regulates spermatogenesis and the growth of the ovarian follicular granulosa cells.

Both sexes have the same steroid hormones, they differ in that men make more testosterone and only 0.25 percent of it is converted to estrogen, whereas women make less testosterone and half of it is converted to estrogen. In females the intermittent production of sex hormones is responsible for the cyclic changes associated with menstruation. Thus, women are fertile only about twelve hours per month whereas men's fertility is constant. Another difference between men and women is that women are born with a limited number of eggs which subsequently undergo apoptosis; the production of sperm is basically unlimited.

Sexual behavior is influenced by gonadal hormones acting on neural systems. It is affected by a wide array of inputs including metabolic balance. Mating strategies have been well characterized in animal studies. In males, the medial preoptic area has receptors for testosterone and the amygdala is involved in motivational behavior. In females the ventromedial hypothalamus, as well as other areas are involved. Of interest, it has been recently discovered that vasopressin and oxytocin, released during mating and childbirth, may facilitate pair-bonding and mother–child bonding, respectively.

Sexual desire or libido is partly under the influence of sex steroids in men but not in women. Sexual activity in humans is initiated by the brain or by mechanoreceptors present in the glans of the penis and in the clitoris and labia. Erection and engorgement are mediated by the parasympathetic nervous system which stimulates the local synthesis of vasodilators such as nitric oxide. After a plateau, the orgasm is triggered by the action of the sympathetic nervous system leading to ejaculation and contraction of the vaginal walls. This is followed by a recovery period.

Pleasure and Pain

Most organisms exhibit specific behavioral responses to pain and pleasure which result in approach or withdrawal from a specific situation. In humans pleasurable stimuli such as sex and food activate the brain reward system which consists of dopaminergic neurons in the ventral tegmental area and its projections to the nucleus accumbens and the prefrontal cortex. Many agents are known to interact with this system and hijack the underlying mechanism. Stimulants such as cocaine and amphetamines interact with dopamine transporters in the brain. Opioid drugs such as morphine or heroin have been found to bind to specific receptors in the brain and are displaced by specific antagonists such as naloxone. Of interest, there are endogenous ligands (enkephalines, endorphines and dynorphins) that bind to these receptors. These endogenous agents act as pain modulators. The brain and other organs have cannabinoid receptors that react with endogenous (anandamide) and exogenous ligands (THC, the active component of marijuana). These agents may also have important effects in the reproductive organs and germ cells. Nicotine is an acetylcholine receptor agonist, whereas alcohol interacts with gamma-aminobutyric acid and glutamate receptors.

The response to painful stimuli is another important behavior. Pain protects us from injury by initiating a reflex that causes us to withdraw from the offending agent and the brain creates an unpleasant sensation that becomes the psychic adjunct of the reflex. The former aspect of pain has a protective effect, whereas the latter is not protective and may be the cause of chronic pain even in the absence of stimuli. The normal relationship between pain and injury is disrupted in neuropathic pain. This type of pain, like phantom limb pain, tends to cause severe emotional, physical and economic stresses.

Nociceptors are sensors that respond to different forms of energy that can produce tissue injury. They exist in the skin, muscle and joints and internal organs. Anatomically they are associated with free nerve endings, A-delta and C fibers, which are the fibers with the slowest conduction velocity of nerve impulses. Some receptors respond only to high-intensity mechanical injury and others respond to a variety of mechanical and chemical stimuli and to changes in temperature. As mentioned, capsaicin, an ingredient of red-hot chili peppers, interacts with the membrane of receptors and opens an ion channel that activates the sensors that send nerve impulses to the brain. The result is an intensive burning sensation similar to that induced by touching a hot object.

Drives and Motivations

After the behavioral responses to pain and pleasure, the next level of life's regulation involves the response to drives and motivations such as hunger, thirst, curiosity, sex and play. The physiologic mechanisms responsible for some of these responses have already been described in previous sections. The triggering factor is usually internal, such as a diminution in energy supply or the increased blood osmolality caused by dehydration. The ensuing behaviors are usually aimed at the environment and include a search for the missing nutrient or water. Some drives are periodic and limited to seasons or physiologic cycles.

Emotions and Feelings

The basic emotions such as joy, sorrow, fear and anger originate in the "limbic system" which comprises several parts of the ancient subcortical regions of the brain. Emotions propel us to engage in behavior patterns designed to protect the organism. They are the substrate of the machinery for producing feelings, a brain map or a mental image that directs the future actions of the organism. Feelings represent the mental idea of the body when it is perturbed by emotions. The conscious awareness of feelings is elaborated in many brain regions including the prefrontal insular and cingular cortices and brain stem nuclei (Damasio, 1999). The cognitively elaborated basic emotions give rise to the social emotions such as shame, envy and jealousy. The hippocampus and other areas are involved in the storage and interpretation of short and long-term memories of emotion-inducing events.

Memory

Activation of emotion-related memories helps to choose among competing options. The brain has the capacity to store information acquired during learning and to retrieve at will. It also has the capacity to forget irrelevant information that we no longer need. Humans have two memory systems: in explicit or declarative memory the material is available to consciousness and can be expressed by language, in implicit memory an acquired skill is performed automatically. The brain structures underlying declarative memory are located in midline diencephalic area and in the temporal lobe, particularly the hippocampus. Non-declarative memories are stored in the basal ganglia, the cerebellum and the pre-motor cortex. Memory can also be categorized into immediate, short-term and long-term memory. Working memory is a type of short-term memory that is used to carry out sequential events. Studies in Aplysia have shown that the release of glutamate is involved in the formation of short-term memories, whereas long-term memory requires synthesis of new proteins and growth of new synaptic connections.

Language and Cognition

Only humans have highly-developed systems, verbal and non-verbal, to communicate with each other and they are the only living organisms that exhibit advanced cognitive functions. Language is lateralized to the left cerebral hemisphere in ninety percent of the population. The cortical areas important for language are Broca's area, Wernicke's area, the primary auditory areas and their close association areas. Lesions in the pre-frontal cortex result in motor or non-fluent aphasia. Damage to the parietal association area may be associated with difficulties in visual and motor integration and constructional tasks and damage to the temporal area may produce agnosias, difficulty in recognizing, identifying and naming different categories of objects.

The prefrontal cortex is responsible for the "executive" functions of the mind such as working memory, planning, attention, emotional regulation, problem-solving and abstract reasoning. The dorsal and lateral regions, specialized in cognitive intelligence, connect with the hippocampus and the left brain. The middle region, specialized in emotional responses, connects with the amygdala and other parts of the limbic system. Damage to the pre-motor cortex may be associated with inability to perform learned skills (apraxia), changes in muscle tone and the appearance of primitive reflexes like sucking and rooting.

Thinking and reasoning are pivotal processes by which we achieve adaptation and mastery of environmental stimuli. Piaget suggested that these characteristics are neither innate nor extracted from the environment; they are constructed in childhood from interaction with the environment. Recent studies suggest two lines of research, one line claims that the capacity of young children was underestimated, and the other line focused on the difficulty encountered by adults in thinking logically. The latter led to the development of dual-process theories that assume that the fast and illogical and the slow and logical thinking compete with each other for control of the human mind.

Development

The human brain undergoes a maturation process that begins *in utero*, continues in childhood and doesn't end until the early twenties, in order to achieve an adult level of functioning. Early development results from the complex interaction between genes and environment and between the child and his caregivers. During this time the individual struggles to develop a distinct and healthy personality that persists over time and is characterized by features that make him a unique individual (Bienenfield 2006; Plante 2006). Personal identity is an ongoing process throughout life; people go through stages that aim towards an indivisible whole. These stages of development have been described by Freud, Klein, Lacan, Mahler, Erikson, and others.

Freud postulated that the child goes through oral, anal, phallic, latency and genital stages of psychosexual development. The healthy ego performs numerous external functions including perception, motor coordination, reality testing, language and cognition. Internal tasks such as control of drives or id, modulation of affect and activation of defense mechanisms are also functions of the ego. Initially the child introjects parental and societal rules and regulations. In the adult these introjected rules help create values, conscience and ego ideal, structures that we associate with the superego. A healthy ego is necessary to moderate the conflicts that often arise between the superego and the id.

Klein described two developmental stages, the depressive and the paranoid-schizoid positions. Lacan proposed three orders of experience through which psychic life becomes organized: the imaginary, the symbolic and the real. The "mirror phase" is the entry point to the imaginary order; the mirror image creates the fiction that the child is whole and has a definite identity.

According to Mahler during development the infant learns to differentiate between the self and the mother. Subsequently, the positive and negative representations of the self and the mother are integrated into a coherent self. This "object constancy" is needed for the child to go through the separation–individuation stage, which is usually complete by the age of thirty months. A secure attachment to the primary caregiver is considered to be a necessary step to achieve individuation and emotional health.

Erikson's psychosocial theory described eight stages of life: infancy, early childhood, pre-school, school age, adolescence, young adulthood, middle adulthood and maturity. The first four stages correspond roughly to Freud's four stages of psychosexual development. There have been many changes in society and in scientific knowledge that merit a modification of some of the original stages suggested by Erikson.

The first stage of life begins with conception, the process of fusion of the sperm and ovum to form the zygote. From this single cell the full grown organism will develop following mechanisms elegantly described by the new science of evolutionary developmental biology. The most critical period in the life of the embryo is gastrulation. This is the phase in which the single-layered blastula turns into a three-layered structure known as the gastrula. The three germ layers are known as the ectoderm, mesoderm and endoderm; they give rise to all the tissues of the organism. After nine months of gestation the baby is born and begins his existence outside the protective environment of the uterus. Many people have suggested that this abrupt change represents a traumatic event that may influence the emotional life of the individual at a future date.

Contemporary psychoanalytical theories reject Freud's drive–defense model and place more emphasis in relationships. The object-relations theory holds that when the mother fails to provide the necessary support for the emerging self, the child may fail to separate and individuate. Many human and animal studies have also shown the importance of the quality of the attachment to the caregiver for the psychological health of the infant.

During the school years the child learns a value system and learns to incorporate the moral, religious and political beliefs of the family and the society. The arrival of adolescence removes the umbrella of parental protection and the vulnerable child may be unable to cope with the new challenges. During this period the individual has to learn to interact with members of the same and the opposite sex. Currently, adolescents have to face problems such as drugs, violence, family disruption, teenage pregnancy and sexually transmitted diseases that were less common in prior generations.

Early adulthood is the time when the individual uses his education and his creativity to contribute to society and secure a steady source of financial support. The other important task of early adult life is to find love and a partner to share life with and perhaps raise children. During middle adulthood many people experience a mid-life crisis. They question their original values, their emotional life, their career choice and their relationships. Many people divorce, change careers or begin to re-examine their political and religious views at that time.

The final stage of life may no longer be associated with retirement because many individuals are forced to work past retirement age or may be capable of continuing to be active in work they enjoy doing. This stage is however associated with the awareness of aging, the development of chronic diseases and the loss of friends and relatives. More importantly, in this stage people reevaluate their lives and their accomplishments, look again for the meaning of life and prepare themselves to face death.

Personality

Normal personality requires the development of healthy and characteristic ways of thinking, feeling, behaving and relating to others. Most people have some sort of understanding of who they are; they develop a self-concept as a discrete and separate individual with its own personality, gender and ethnicity and with its own religious, cultural, political and economic views. Personality includes temperament, the individual differences in arousal and mood that are influenced by hereditary factors, and character which depends on the environment.

The person's character is shaped by developmental processes and life experiences, especially the interaction with the primary caregivers. When the caregiver is attuned to his emotional needs, the child feels secure and develops a positive internal representation of the self and the other. If his feelings are validated and he comes to know himself as being known by another he develops a well-integrated self and the capacity to read the mind of others.

The five-factor model is the most widely used classification of personality traits (Silverstein, 2006). Individuals with a low extraversion score tend to be quiet and shy; however, in contrast to schizoid individuals, they do not enjoy being alone and are not indifferent to criticism. A high score in neuroticism is found in individuals who are anxious, insecure, self-conscious, vulnerable, and tend to avoid social interactions. A high score in conscientiousness is typical of individuals who prefer competence, order, achievement and self-discipline. Some people work very hard, though, not because they like what they are doing, but for fear of making mistakes or being criticized. A high score in openness identifies individual that are creative and open to new experiences. Individuals with low agreeableness score tend to be antagonistic, suspicious and deceptive. Some experts add a sixth factor, honesty/humility; a high score describes individuals who are honest, sincere and ethical.

Personality disorders are characterized by defective self-integration and the resultant problems with intimacy and empathy that interfere with love and work, the main tasks of adult life. Some psychologists claim that these disorders represent maladaptive variants of personality traits that are present in all individuals. Most experts, though, claim that they are caused by a combination of genetic and environmental factors such as abuse or neglect. The latter may induce epigenetic changes that could be transmitted to the progeny.

There are many types of personality disorders. They defy categorization because insufficient knowledge about psychodynamic and biological etiologic factors. Clinical descriptions include patients that are helpless, incompetent, vulnerable, glamorous, paranoid, avoidant, or compulsive. Affected individuals may not be able to know how they feel (alexithymia) and may be unable to attend to the mental states of others (mentalization). Their problems with interpersonal relations include lack of empathy, trust and connection.

Narcissistic personality disorders are difficult to recognize because we live in a narcissistic culture and are consumed by the glamour of celebrities, money and beauty. We are enthralled by the electronic media that thrives on superficial images and ignores substance and depth. We view unbridled consumption as the road to happiness and self-interest as the basis of our economic growth.

Individuals with full-blown narcissistic personality disorder crave admiration, lack empathy, feel entitled and are exploitative. They may use other people to maintain self-cohesion and as source of gratification. Closet narcissists are timid, self-effacing and focus on others, always looking for evidence of slights or criticism. Some men become womanizers and keep changing partners when their needs are not met. Psychopaths lack empathy and are manipulative and deceitful.

Individuals with borderline personality disorder feel empty, can't be alone and experience fear of abandonment and emotional instability (Meares, 2012). Affected individuals may experience lack of autonomy, impaired cognitive skills and identity diffusion, which consists of contradictory character traits, temporal discontinuity, body image disturbances and lack of authenticity. They often utilize primitive defense mechanisms such as splitting and projective identification.

Failure to obtain the developmental support needed to individuate sometimes leads to a less severe identity disorder. The affected individual may become a "chameleon" and adopt the identity of those around him. His boundaries may become porous and he may not develop a distinct moral, social or political viewpoint. Some individuals fit in the middle of a continuum between healthy and pathological identity. Although their personal identity is not overtly fragmented, they exhibit what is called "other identity"; they need to identify with something other than themselves in order to support their self-concept. These individuals have a need to dress in a certain way, to drive a particular brand of car, to belong to a particular group or to live in a particular neighborhood in order to have a firm identity.

CHAPTER 8. HUMAN NATURE

Most religions affirm that we have a God-given nature that determines our behavior and gives purpose to human life and human society. There are some philosophers who have selected a number of fixed characteristics that most humans exhibit. Others, like Sartre, deny the existence of human nature and claim that we have to create our own reality. Marx held that our natures are not fixed but are influenced by the particular economic stage of our society. Most naturalists, though, believe that that there is such thing as "human nature" (Harris, 2011). Darwin affirmed that we are a product of evolution with biologically determined, species-specific behaviors but have highly developed brains that may help us to modify our innate dispositions. We are part of nature but transcend it by being endowed with reason and self- awareness.

For many years there has been a controversy concerning which factor, genes or the environment, has the greatest impact on human nature. Sociobiologists, and more recently, evolutionary psychologists, have argued that human nature is influenced mainly by the genes. On the other hand, social scientists played down the role of genetics and concentrated primarily on the role of the environmental influences such as culture and conditioning. This model (the blank-slate) posits that human nature is very plastic: we are not born with specific traits, and social hierarchies are not innate, they are a social construction and human beings can be taught otherwise (Pinker, 2002).

Developments in linguistics followed by the growth of sociobiology, the application of evolutionary principles to explain social behavior (Wilson, 1998), and evolutionary psychology, the study of the role of evolution in the development of our psychological traits (Barkow, Cosmides and Tooby, 1992), weakened considerably the position of proponents of the blank-slate view of human nature. Evolutionary psychology is based on the premise that behavior is driven by adaptations. An adaptation is a characteristic that has arisen by natural or sexual selection and that helped solve the problems of survival and reproduction in a given species. Evolutionary psychology has been challenged recently by systems theory and by the epigenetic paradigm; the consensus now appears to be that our behavior depends on the interaction between genetic and environmental factors, and that our genome can be reconfigured by the environment.

Evolutionary psychology distinguishes between proximal causes of behavior, such as environmental factors, and ultimate causes, which reflect the overarching evolutionary goal of fitness or the perpetuation of the genes. For example, men typically tend to seek financial assets, status and resources, and women pursue beauty, health and youthful appearance but their ultimate goal may be finding a mate and the perpetuation of their genes. Some of our behaviors are not always adaptive for modern man, such as our penchant for fats and sweets, which were adaptive when high caloric food was scarce.

Humans like all living organisms are a product of evolution, our roots are in nature. We can trace our lineage back billions of years, to primitive unicellular organisms, not to the mythical ancestors described by Scripture. In spite of the variety of organisms produced by evolution, all living things, including humans, are made of cells. Cells are biological machines comprised of thousands of proteins that carry out precise biochemical functions and which are themselves coded by DNA. Energy for the cellular processes comes from the metabolism of organic compounds such as glucose.

Like all living organisms, we display the basic behaviors that are essential for the survival of the individual and the species. These basic behaviors include identifying friend and foe, establishing and defending status and territory and being able to reproduce. To accomplish these tasks most living organisms share not only the cellular structure, but also the same energy currency (ATP) and the same mechanism of heredity (DNA). Except for the most primitive multicellular animals, the behavior of living organisms is regulated by a specialized cell type, the neuron, which is capable to generate action potentials and to rapidly transmit them to connecting cells.

Behavior is the direct result of neuronal activity in the nervous system. Some invertebrates have fixed-action patterns of response, whereas more complex organisms have complex brains and the capacity to modulate their responses based on experience. As mentioned before, human behavior reflects the activity of the motor nervous system which is itself influenced by inputs from the sensory and cognitive neuronal systems and the basic background activity of the nervous system. The external and internal environments provide sensory feedback signals, direct sensory inputs to the motor system. The background activity of the nervous system, including the circadian or day–night cycles, is controlled by specialized neuronal groups that in humans extend from the brain stem and hypothalamus to the cerebral hemispheres. The cognitive system is responsible for knowing, understanding and planning actions.

The human brain consists of complex neural networks, shaped over millions of years of evolution. The overall structure of the brain is genetically hardwired; however, the neural circuits undergo extensive sculpting and rewiring in response to a variety of environmental stimuli. Our social behaviors can be better understood in the light of natural selection and our adaptations to the hunter-gatherer lifestyle that was dominant during most of our evolutionary history. These adaptations include ways to find food and water, identify and fight predators, avoid threats, bond to friends and family and find and retain mates. Evolutionary psychology explains why these behaviors are a recurrent feature of human societies and why they include competitive as well as cooperative behaviors. It posits that the human mind is modular; it has a large number of semi-independent modules that are responsible for these behaviors.

Humans evolved to detect and minimize threats to their social connections. They follow the social organization of some non-human primates. Beginning about fifty million years ago primates transitioned from solitary life to loose aggregations of males and females. Orangutans are still solitary animals, whereas chimpanzees and baboons live in large groups that include males and females. Both pair-bonded individuals and one-male harems are found in primates. The former can be observed in gibbons and humans and the latter in gorillas. In pair-bonded animals promiscuity still occurs, however. It has an obvious advantage to males in that other males will be raising their offspring. Promiscuous females may benefit from the improvement in the genetic makeup of their descendants.

Humans are warm-blooded animals and like all mammals they carry their offspring in the womb. We nurture helpless newborns with immature brains for prolonged periods of time, which facilitates learning and exploration but requires a lot of care and protection. During this period the offspring forms close connections with the parents. The ability to imitate, mediated by mirror neurons, is present in the great apes which are also able to learn sign language. There are traits that are uniquely human, however, including language, the capacity to read the minds of others, the sense of self and a self-control system that ensures that we follow social norms and values.

Our Dual Nature

Common sense predicts that human evolution would select selfishness and self-preservation as desirable traits to be passed to the progeny. Nonetheless, evolutionary theory suggests that being unselfish and exhibiting sociality and a moral disposition may also have a beneficial effect on survival. We have the potential to behave selfishly or altruistically, both dispositions may have an evolutionary origin.

Cooperation and the need to connect with others are widespread in nature. For instance, single cells go into symbiotic relationship with other cells and can form aggregates that eventually become multicellular life forms exhibiting a division of labor and specialization. Cooperation reached a high level of complexity in the social insects. Most mammals also exhibit elaborate pro-social behaviors. For example, some monkeys are known to emit specific warning calls in the presence of predators, a behavior that may compromise their own safety.

We help others that are genetically related (kin altruism), because that is a way to perpetuate our own genes. We may cooperate with others to receive payback for a favor or for them to create a protective environment for the group (reciprocal altruism). Social support is so critical that evolution has given us additional neural networks and mechanisms to connect to others. The ability to recognize social pain and the need for others, and reacting accordingly, has been an important determinant in the evolution of humankind. Sociality is present in our animal ancestors in which mirror neurons are responsible for empathy and neuropeptides, like oxytocin, help cement the bonds between members of the group (De Waal, 2013). Pro-social behavior has been reinforced by our capacity of mindreading (mentalizing) and by the incorporation of socially-accepted ethical rules.

Love and caring are prevalent in the world, but so are aggressive and competitive behaviors. Aggression is commonly displayed by living organisms, including humans, seeking to obtain resources, status or mates in order to survive and reproduce. Since the beginning of recorded history human society has often failed to tame basic competitive and aggressive behaviors that were acquired during a long evolutionary process; cultural and religious ethical rules have been unable to trump biological factors.

Many people don't understand our dual nature and wonder about the origin of our bad behaviors. Our bad behaviors, like those of all living organisms, are often motivated by the need to survive and reproduce. Most animals, especially vertebrates, have evolved strategies to deal with the four fundamental problems of life: identity, reproduction, hierarchy and territoriality. The basic communicative displays triggered by these problems, signature, courtship, challenge and territorial, are more elaborate in human societies; they manifest themselves as a struggle for social equality, reproduction, dominance and financial success.

During evolution the nervous system assumed control of the homeostatic functions necessary to deal with life's problems. Stimuli from the environment get transformed into action potentials that travel rapidly to the brain via peripheral nerve fibers. In the brain this information is processed into things that the animal can see, hear, smell, taste and touch. The information may lead to specific behaviors that are carried out by other components of the neural network. Damasio claims that the control systems underlying these functions are arranged following the nesting principle; the simple ones are incorporated into more elaborate ones forming a tree-shaped structure.

The controversial triune brain model posits three stages of brain evolution. The so-called reptilian brain consists in structures that comprise a substantial portion of the forebrain of reptiles; the mammalian brain achieves a highly complex level of differentiation in the mammals' limbic system and the neo-mammalian brain, which is jury-rigged from pre-existing structures, is characterized by the marked growth of the cerebral cortex.

Our reptilian brain oversees the most basic and automatic processes necessary to maintain life. The latter include mechanisms for maintaining arousal, body temperature, metabolism and the constant composition of body fluids. The distribution of oxygen and nutrients by the circulation assures the viability of peripheral tissues and the availability of substrates needed to maintain the structure and function of the different organs. Basic reflexes guide organisms to move away from painful and noxious stimuli and to seek safe and pleasurable experiences. The immune system is the first line of defense against invading microorganisms.

In the next level of life regulation are the drives and the emotions. Sensory input may reach the limbic system directly and elicit a fast response; sometimes it also travels to the sensory cortex in order to determine its emotional significance and the appropriate response. We share with other mammals the physical expression of emotion. Most experts agree that there are four sets of opposing basic emotions: joy/sadness, fear/ anger, acceptance/ disgust and anticipation/ surprise. Secondary emotions, like shame and jealousy, are formed by combinations of two or more basic emotions; they help us to navigate the social world. The conscious awareness of emotion, in combination with memory and other higher functions, provides the necessary information to guide subsequent behaviors.

The highest level of brain functioning resides in the cerebral cortex. The prefrontal cortex in particular is essential for emotional regulation and the brain's executive functions which include working memory, planning, attention, decision-making, problem solving and abstract reasoning.

Humans have the capacity to communicate verbally and non-verbally with others. Language is lateralized to the left cerebral hemisphere in most people. Consciousness, the awareness of the self and others, and the capacity to read the mind of others are also highly developed in humans.

The way our brains have evolved may explain our pro-social and anti-social behaviors and the reason for the conflicts between reason and emotion. Our prefrontal cortex may fail to regulate the limbic system, emotion sometimes trumps reason. We frequently make fast and intuitive decisions without rational planning. We tend to avoid incompatible views about ourselves, cognitive dissonance, and suppress negative self-assessments. We often deceive others or ourselves in order to gain favors or acquire financial or social advantages. We may use the confirmation bias to cherry pick things we want to believe in and ignore other views. We tend to resist the idea that random events exist and tend to find patterns in nature that don't exist. We are often in denial of unpleasant events including the reality of death and suffering.

Humans tend to stereotype others, perhaps to avoid potential predators and this often leads to prejudice. We may fear anonymity and lack of identity and identify with celebrities or material possessions. Many people suffer from boredom and lack of direction in life and may try to find escape from boredom in loud music, violent sports or salacious movies. We deal with existential anxiety by activating our brain reward system with food, sex or drugs and often become addicted to these agents. We fear the unknown and search for meaning and values and often cling to superstitions or believe in supernatural entities.

Because the way we evolved, we tend to exhibit conflicting dispositions. Ethologists distinguish between hedonic and agonic forms of social organization. The former is based on our desire for cooperation, equality, and altruism, whereas the latter is related to our competitive, ethnocentric and hierarchically nature. Most of us exhibit features of both forms of social organization. Thus we may cooperative at one time and competitive at other; we may be averse to inequality, yet in our daily life we exhibit prejudices and act in our self-interest.

Many of our moral, religious and political views are determined by personality traits or biological factors, not by culture or conscious deliberation. These traits have their roots in behaviors that develop in all living organisms in order to survive and reproduce; they are difficult to modify or eradicate. Some of our personality traits reflect our tendency toward tribalism, our degree of tolerance to inequality and our perception of human nature.

According to Tuschman we have different ways to react to tribalism and its three components: ethnocentricity, religiosity and sexuality. These traits may have originated from the need to achieve a healthy progeny by finding an optimal balance between inbreeding and outbreeding. The more ethnocentric, religious and sexually intolerant people are, the more likely they are to mate with members of their own in-group. Thus conservative individuals favor xenophobia over xenophilia, religiosity over secularism and intolerance to non-reproductive sexuality over sexual freedom.

People also tend to exhibit different degrees of tolerance to inequality. These traits may have their evolutionary roots in behaviors designed to cope with conflicts within the nuclear family and to deal with parent–sibling conflict and sibling rivalry. Conservatives usually favor hierarchies, whereas liberals are more egalitarian (Meyer, 2007). The former tend to attribute the status of high-ranking people to their superior faculties, work ethics or moral qualities.

Humans also differ in their perceptions about human nature. Some individuals think that humans are basically competitive and aggressive and others believe that we are cooperative and empathic. These traits may have their evolutionary roots in the need to balance altruism against self-interest. Liberals think that we are basically cooperative and therefore they tend to trust others and establish alliances with out-groups. Conservatives believe that we are basically competitive and favor self-interest.

Religion tends to create the image of a powerful God that will protect us in a dangerous world and will help us to curb our bad behaviors. Humans may be wired to believe in myths, superstition and supernatural beings (Shermer, 2011). These belief systems favor survival because they promote group cohesion. Nevertheless, religious groups may stereotype and act violently against members of other faiths; religion is one of the most important causes of human conflict. Religions carry the message that we should avoid greed and selfishness and love our neighbor as much as we love ourselves. Nonetheless, these prescriptions have often failed to change biologically-determined selfish human behavior.

We claim to be autonomous, self-determining individuals and ignore the fact that our unconscious biological dispositions often trump rationality and play an important role in shaping our behavior and decision-making (Kahneman, 2011). We are able to reason and to have self-awareness but at the same time exhibit uncontrolled emotions. We are prone to addictions and deception, we also engage in self-deception: we deny the truth to ourselves, repress painful memories, rationalize immoral behavior and boost our self-opinion.

Our advanced civilization has created numerous scientific and technological advances including computers, spaceflight, nuclear energy, and sublime works of arts and literature. On the other hand, the way that we humans have behaved throughout history suggests that our higher functions and pro-social emotions have often been unable to tame our competitive and selfish drives and emotions. We are extremely aggressive and irrational beings. We are prone to self-destruction and to exert violence against other living things and the environment.

Human Violence

In addition to the violence and cruelty found in nature, humans often exhibit remarkable aggressive behaviors. They evolved to help humans survive in a hostile world and compete for status, food, territory and mates. A few thousand years of civilization have not been able to erase behaviors that arose during millions of years of evolution. Social factors such as oppression, religious differences or inequality frequently trigger aggression. Thus individuals who are usually cooperative and caring may engage in violent acts against others or themselves ignoring the teachings of religion and society. Nations engage in peaceful trade and cultural exchange but still make war in spite of the loss of human lives and national resources.

Interpersonal aggression is usually reactive, induced by hatred and caused by actual or perceived provocation or projection of unwanted traits. It may be physical or verbal and often is indirect, like destroying someone else's property or reputation. Biologic or structural factors may be at the root of non-reactive or instrumental aggression. Freud believed that sex and aggression are our basic motivations; he believed that the latter is a major threat to civilization. Buss studied murder rates in young men and concluded that there is a connection between reproductive competition and violence.

External conditions such as humiliation, crowding, poverty, noise and heat may breed violent behaviors; they may create a violent "culture of honor" that helps individuals to gain power and self-respect. The studies of Milgram and Zimbardo suggest that evil acts are not the result of the individual's basic disposition but of coercive forces that propel human choice. The "Lucifer effect" explains how peer-pressure and other situational conditions may lead to evil acts even in normal and ordinary people.

Most violent offenders are men. Despite the evidence that testosterone, the male hormone, facilitates aggression in many vertebrate species, its role in humans is still being debated. Male aggression against females in the form of wife-beating and sexual violence is a major social problem. Some of the abusers have been themselves victims of family violence or suffer from emotional disorders.

Damage to the empathy circuit in the brain, which includes the amygdala, the caudal anterior cingulate cortex, the anterior insula and the pre-frontal cortex, may lead to aggressive behavior in some people. The right temporoparietal junction helps us to imagine someone else's thoughts and the superior temporal sulcus monitors the direction of gaze and emotional reactions. Bilateral temporal lobe lesions may cause violent behaviors and brain scans from aggressive individuals have shown abnormalities in the empathy circuit. Experimental studies have shown that damage to certain areas of the pre-frontal cortex may also elicit aggressive behavior. Genetic factors may cause low empathy levels and predispose to violence. Some scientists believe that the principal neurotransmitter involved in aggression is serotonin. Serotonin exerts an inhibitory effect on aggression and aggressive individuals have been shown to have low blood levels. Decreased blood levels of monoamine oxidase (MAO), the enzyme that degrades norepinephrine, may also cause increased aggression. Genetic factors may also influence how our brains react to emotions and may predispose to autism and personality disorders.

Violence may decrease by limiting the widespread sale of handguns, regulating the release of violent movies and video games and perhaps by abolishing state-controlled violence such as the death penalty. Violence can be curbed in the home by education and therapy designed to prevent marital, child and elderly abuse. Perpetrators should not bear full legal responsibility for their acts if they are suffering from emotional disorders or cerebral dysfunction. Society should implement a more effective system of law and order, reward cooperative and altruistic behaviors, and create a more stable social system. Perhaps in the future we may be able to identify and correct the biological causes of aggression.

War is one of the most destructive manifestations of collective aggression. It has resulted in much pain, suffering and economic loss and now threatens survival of the planet. War was defined by Clausewitz as an act of violence intended to compel our opponents to fulfill our will. There are three general philosophical theories about war. Proponents of the just war theory claim that for war to be justified it has to be prompted by a just reason; for example, self-defense or protection of innocents. Realists question the need to use moral concepts to justify war. They claim that power and security issues and the need to maximize self-interest may be reasons to go to war. Pacifists oppose war as a means to settle disputes. Principled pacifists find no moral justification for killing. Pragmatic pacifists feel that the benefits never outweigh the costs of war but they allow it under certain circumstances. Pacifists have been criticized for lacking realism, rewarding aggression and failing to protect their own.

The causes of warfare have been classified into biological and environmental. Ethologists have explained human aggression on the basis of behavioral biology. Emphasizing the place of human beings within the animal kingdom they postulate that aggression is triggered by external events in individuals with pent up aggressive energy. Evolutionary psychology claims that behavioral modules designed to foster intergroup aggression may lurk behind our long history of warfare. Freud in his later writings suggested that we have an instinct tending to preserve life and an aggressive and destructive death instinct. Other studies also support a contribution of genetic factors to aggression.

There are many environmental causes of war. Liberals claim that wars are waged for predominantly political reasons, whereas socialists stress the role of socioeconomic factors. The latter claim that wars may be started by states seeking raw materials, new markets or cheap labor. Most wars have been initiated to obtain resources and territory, to defend against expected aggression, because of religious or political conflicts or because of the need to protect minority groups. Extreme nationalism, allegiance of men to the nation and the nation to the state, and the work of special interest groups, may also be responsible for the initiation of war.

The history of humanity documents the high frequency of warfare and the extent of the atrocities and unspeakable acts of cruelty performed by human beings. The beginning of the agricultural revolution often created intergroup conflicts, our ancestors engaged in continuous attacks and pillage of neighboring tribes in order to obtain accumulated resources. After settling down in villages and cities, ancient civilizations developed professional armies for their protection and as a way to gain additional resources and territory.

Toward the end of the third millennium BCE, Sargon had conquered most of Mesopotamia. Subsequently, the Babylonians under Hammurabi became the leading power in the region and fought several wars against the surrounding countries. In Egypt, during the New Kingdom, military conquests created an empire stretching from the Euphrates in the north to Nubia in the south. It reached its pinnacle of wealth and power under Ramses II who utilized chariots to battle against his neighbors. Egypt's territorial expansion came to an end after the battle of Kadesh against the Hittites.

At the beginning of China's Bronze Age, the Shang dynasty expanded its territory to the Yangtze Valley, but in the eleventh century BCE they were conquered by the Chou, of different ethnic origin. In India, the Indus civilization, like those of Egypt and Mesopotamia, was based on flood-plain agriculture. It ended when the Aryan invasion from the northwest conquered the cities of the dark-skinned indigenous people. The Minoan civilization in Crete achieved considerable development until it was overtaken by Mycenaens from the mainland. The Trojan wars were related in the Homeric poems. These poems probably represent a fictional account of the warfare between the Hittites and the Mycenaens.

In the centuries after 1000 BCE, the Iron Age Civilization spread throughout the world. Iron technology made it possible to increase in productivity of these societies and to improve the construction of weapons. After the Bronze Age, Assyria became the dominant power in the Near East. In 689 BCE, Sennacherib, King of Assyria, captured and destroyed Babylon, slaughtered its inhabitants and diverted rivers to flood the city. Ashurbanipal led the invasion of Egypt as far south as Thebes. After his death, however, the Assyrian power declined and its capital Nineveh was sacked and came under the control of Babylon.

Nebuchadnezzar drove away the Egyptians and captured Jerusalem, driving its inhabitants into exile. The great Persian leaders Cyrus, Cambyses and Darius created a large empire. In 539 BCE, Babylon fell to the Persians under Cyrus I. Over the next several centuries, the Persian Empire was governed successively by the Achaemenid, Arsacid and Sasanid ruling families. The first ruling family came to an end after the invasion by Alexander the Great in 334 BCE and the last victories by Arabs at Al Qadisiyya and Nihavand in 637 CE and 642 CE, respectively.

The Greek city-state of Athens perhaps remains best-known for its pioneering use of elements of democracy in its political functioning. The Persians mounted several military expeditions against the Greeks. They were rejected and lost major land (Marathon) and naval (Salamis) battles. The Peloponnesian war against Sparta resulted in the collapse of the Athenian Empire. In 338 BCE, the Macedonians became the leading power in Greece. The Greek phalanxes under Alexander the Great fought and conquered the Persian Empire. In 330 BCE, they plundered and burned the palace of Xerxes. At Granicus, Alexander massacred thousands of enemy troops. His empire extended from Egypt to the mouth of the Indus River.

The expanding power of Rome resulted in military conflicts against Greece and Carthage. Pyrrhus was one of the Greek generals who fought in Sicily and southern Italy against the Romans. He won the bloody battle of Asculum in 279 BCE with crippling losses (thus the term Pyrrhic victory, a victory in which the winner suffers so much that the victory was hardly worth winning). The expansion of the Roman power led to the Punic wars and the defeat of the other major power in the western Mediterranean, Carthage. The First Punic War drove the Carthaginians out of Sicily. In the Second Punic War, Hannibal brought his army and elephants over the Alps into Italy. After several defeats, the Romans were able to drive Hannibal out of Italy and defeat him in the battle of Zama in North Africa.

The period from 133 BCE saw an increasing turmoil associated with Rome's continuous expansion. Among the consequences of military victories abroad there was an influx of goods, arts and people into Italy. The conquest of the Greek empire brought important changes on the political, social and cultural life of Rome. Greek works of art and Greek literature, science and philosophy had a profound effect. Wars also provided slaves and other workers, leading to a rapid growth of the urban population. In 49 BCE Caesar crossed the Rubicon and became the sole dictator of the emerging empire. In 31 BCE, his grand-nephew Octavian, the future Emperor Augustus, won the battle of Actium against Mark Antony and Cleopatra, and became the ruler of the Roman Empire. Augustus introduced reforms that were far-reaching. He reorganized the Senate and the army and made the Emperor the focus of all Roman religious ritual. These reforms brought Rome two and a half centuries of almost uninterrupted peace and prosperity.

As the Roman Empire expanded and organized, it became the foundation for the development of the Mediterranean world in the first three centuries of the Common Era. The emperor's figure was central to the empire; statues and coins were constant reminders of his presence. In the succession of emperors, many were assassinated and others, like Nero, committed suicide. Later in the second century, pressure began to mount in the frontiers of the state. Marcus Aurelius spent most of his reign at war with barbarian invaders and his successors faced threats from the north and from the Persian Empire under the Sassanids.

During the rule of Diocletian and his successors, the emperors owed their position to the military, not to the Senate. In 330 CE Constantinople rose in the Eastern portion of the empire and became its capital after the death of Theodosius in 395 CE. In the fifth century, the Western empire was dissolved in the face of continuous incursions by barbarians, many of whom converted to Christianity. The eastern or Byzantine empire was Greek-speaking, culturally diverse, and Christian. It lasted until 1453, when it was overrun by the Ottoman Turks.

The migration (hijra) of Mohammed to Medina in 622 CE marks the beginning of the Islamic era. His successor, Abu Bakr, completed the conquest of Arabia and entered southern Palestine. Subsequently, Umar ibn al-Khattab took Damascus from the Byzantines and led invasions into Mesopotamia and Asia Minor. In 643 CE, Persia was overrun, and subsequently the invaders extended their conquest to occupy Kabul and Sind in northern India. In the west, the Arabs invaded Egypt and northern Africa; in 711 CE, they crossed the Straits of Gibraltar and conquered the southern part of Spain (Al-Andalus). Spain remained under Muslim occupation until 1492, the year of the discovery of the Americas by Columbus. In Southeast Asia and Africa, many converted to Islam due to the influence of traders and preachers.

The Crusades were an attempt by Christians to recover parts of the "Holy Land" from Islamic domination. They took place between the eleventh and the fifteenth centuries, at a time of increasing religious fervor and conversions to Christianity. Support by the popes and their promise of eternal salvation (indulgences) contributed to the enthusiasm of Christians to wage holy war. During the conquest of Jerusalem in the First Crusade, the massacre was of such ferocity that it shocked many of the participant warriors. The Third Crusade was waged in response to Saladin's re-conquest of territories in the Levant including Jerusalem. This crusade was compromised by the bloody capture of Constantinople when the planned campaign in the Holy Land was diverted and turned against fellow Christians. The crusading movement continued with other incursions that took place until the middle of the fifteenth century.

The Inquisition was responsible for the torture and death of thousands of people. It was established by the popes to repress heresy following the Albigensian Crusade. The Spanish Inquisition was created in 1478 to deal with non-believers; Jews and Muslims were forced to convert to Christianity or faced execution. During its initial period, those conducting the Inquisition were responsible for the arrest, torture and execution of a large proportion of those prosecuted. In France, the Protestants (Huguenots) suffered persecution from the Catholic majority. Many years of hostilities followed, culminating in the Saint Bartholomew's Day Massacre that started in Paris and later on extended to other cities.

Human violence has continued unabated until the present. The twentieth century was particularly violent. It is estimated that nine to ten million people died in the First World War and fifty to sixty million died during the Second World War.

CHAPTER 9. MORALITY AND FREE WILL

Morality

Morality addresses how people conduct themselves in society, whereas ethics refers to the branch of philosophy that evaluates questions of morality. Meta-ethics studies the origin and meaning of ethical concepts. It is unclear whether or not there are absolute moral truths. Moral skepticism or anti-realism is the view that morality has no objective or independent foundation. Some individuals think that there are no moral facts, whereas others hold that there are moral facts but they are either subjective or dictated by the culture. Moral relativists claim that there are no absolute moral rules; there are no universal moral principles and morality differs in every society (Wright, 1994). Emotivism claims that ethical statements don't express any facts; they only express the speaker's emotion. Critics argue that if this theory were true then all moral arguments would be impossible. The alternative to moral skepticism is moral realism, the view that morality has an objective foundation. Proponents of this view argue that there are moral facts and that they are independent of our subjective states or the conventions of society. Naturalistic ethics is based on the assumption that ethical judgments follow directly from scientifically discoverable facts. Nonetheless, proponents of the Naturalistic Fallacy, such as Moore, suggest that you cannot derive "ought" from "is." They hold that you cannot move from facts to values. Others, like Sartre, have argued that it is a mistake to assume that there is such a thing as human nature.

The leading normative theories are consequentialist (teleological), duty-based (deontological) and virtue ethics. Consequentialists affirm that the right thing to do is whatever has the best consequences. There are several types of teleologists: the egoist puts the personal self-interest first, the altruist holds that the right thing to do is whatever has the best consequences for others and the utilitarian seeks the greatest amount of happiness for the greatest number of people. The problem with utilitarianism is that is difficult to measure happiness and to compare the happiness of different people. Another problem of this theory is that it could justify actions which are usually thought to be immoral. For instance, it could justify killing an innocent person if it has the benefit of reducing violent crime by acting as a deterrent. Some people embrace negative utilitarianism, they claim that avoiding pain and suffering is more important goal than bringing happiness. Others combine utilitarianism with deontological ethics.

Deontologists follows absolute rules and hold that whether an action is right or wrong has nothing or very little to do with its consequences. Duty-based theories include Kantian and Christian ethics. Kant believed that as rational human beings we have certain duties. These duties are absolute and unconditional (categorical), and apply regardless of the consequences that may follow from obeying them. The underlying principles should be universal and should propel us to treat other people as ends in themselves, never as a means to an end. The problem with Kantian ethics is that it doesn't give practical solutions to many moral dilemmas, especially when it involves conflicting duties. For example, the duty to tell the truth and the duty to protect a relative may conflict with each other when we have to lie in order to protect the life of that individual. Deontology also dismisses human emotions as irrelevant and ignores the consequences of our actions.

Virtue ethics is based on the teachings of Aristotle. It is based on the character of the agent, not in the evaluation of the right or wrong of a particular action. Only by cultivating virtues can human beings flourish. Major problems with this theory are that it assumes that there is such a thing as human nature and that it is difficult to establish which actions count as virtue. Recently, care ethics and feminist ethics, which are based on our capacity for empathy and caring for others, are attracting the interest of many ethicists.

Our conception of what morality is has been shaped by religion. Some people believe that morality can only exist if it is supported by the unchanging and transcendental rules established by religion. Christian ethical theory teaches that what makes an action right is that God commands it and what makes it wrong is that God forbids it. Plato in his dialogue Euthyphro questioned the divine command theory of ethics by raising the following paradox: Does God establishes that something is good because it is good, or is it good because God says so? Religion doesn't always give us a solid foundation for morality: different religions have contradictory moral imperatives and sometimes they are the cause of evil actions. Doing good to expect a reward and to avoid doing evil to escape punishment doesn't appear to be a good moral system. The implication that humans carry the original sin, and thus are capable of evil, has been rejected by naturalists and many theologians.

Humanists have a consequentialist ethical system and claim that morality can exist without religion. Secular ethics is humanistic and naturalistic, firmly grounded in the characteristics of human beings. Darwin postulated that morality in humans probably arose from our ability to evaluate our actions and modify our biologically-determined animal behaviors. Dawkins suggested that evolution has hardwired humans to behave ethically; pro-social behaviors, like empathy and bonding, that are the product of evolution, some of which we share with other primates (Hauser, 2006 and James, 2011). Natural selection would be expected to favor groups and individuals that exhibit pro-social behaviors, any trait that leads to a better adaptation gives an organism a better chance to survive and reproduce. These traits have spread to the population and have helped to challenge the idea that perpetuation of our own genes is the ultimate source of behavior.

Like Dawkins some recent authors claim that what we call "moral behavior" is biologically rooted. Studies in primates by De Waal have documented how pro-social behaviors can be observed in non-human primates. Nielsen suggested that the goal of a secular humanistic ethics is to promote and maximize happiness. Kurtz agrees with Nielson that a naturalistic ethic is unable to make categorical claims. He believes that we recognize objective ethical values but that our intelligent minds are flexible enough to consider extenuating circumstances (Kurtz, 2007; Nielson, 1990).

In the animal kingdom, animals ruthlessly prey on each other. It is not that evil, a mysterious force that emanates from demons, is the way nature operates. To survive, humans and animals have to eat other living organisms. Darwin was appalled by the cruel design of nature. Humans sometimes exhibit aggressive behaviors toward others, especially if they are different from them or belong to a different group. The existence of base and cruel human passions is undeniable; there is no substantial evidence that religion or society have curbed them.

Some individuals choose wicked actions because they see no easy or legitimate way to reach their goals. Other perpetrators are idealists or see themselves as victims. Some people question free will and think that some offenders are not fully responsible for their actions. For instance, there are individuals who lack empathy or don't have absolute control over their passions. A minority are sociopaths, a personality disorder diagnosed in people that enjoy inflicting pain on others without suffering any remorse. There are genetic and neurological components to criminal behaviors, twin studies have shown that one half of the variance in aggression and antisocial behavior is attributable to genetic factors. Expression of these behaviors may require the additional effect of environmental factors such as brain damage and psychological abuse or neglect. Evil-doers often use self-justification to avoid cognitive dissonance, the state of tension that occurs when individuals hold two inconsistent views about themselves.

Questions about the right of animals and vegetarianism, based on moral grounds, have surfaced in the last few years. Most animals are capable of feeling pain and, like us, have a central nervous system. Many large-scale farming operations keep animals in crowded and unsanitary conditions and treat them badly. Bentham claimed that the suffering of animals should be included in any moral assessment of how we should behave. Singer has argued that treating animals' interests different from those of human beings constitutes "speciesism," a form of prejudice and discrimination. Support for "speciesism" is grounded in the Old Testament teaching that God gave human beings dominion over animals and that they exist only for the benefit of humanity.

Some individuals assert that animals don't show respect for each other and that we are justified in favoring our own species relative to other animals. Most non-human animals, particularly mammals, have interests and should be allowed to live their natural life free of the suffering caused by humans. Some people think that we should be vegetarians or vegans and refrain from most forms of animal experimentation. Others claim the animal rights should be recognized by the law. Critics claim that animals cannot understand rights and the associated duties. Animals may not be sufficiently developed morally and psychologically to merit the possession of rights. Other people claim that our duty to treat animals fairly is not a duty to animals but rather to other human beings.

Free Will

Proponents of free will argue that we are free to choose, whereas determinists hold that free will doesn't exist and that events can only unfold in one way. It seems to us that what we do moment to moment is determined by conscious decisions that we freely make. Without free will there could be no moral responsibility. Both religion and the legal system presume that we have free will and are responsible for our behavior. Unfortunately, we don't have enough knowledge of neuroscience to show that free will exists. We don't know that when we make our own decisions nothing causes us to choose the way we do. Some people that believe we have free will claim that their view is supported by the findings of quantum mechanics. Quantum events, however, take place at orders of magnitude several times smaller than those associated with neuronal activity. These events are also random and cannot be correlated with the free will of the individual.

There are some important arguments against free will; if our decisions are caused by prior events, they are predetermined and if they are not determined by prior events they are random. These arguments have been criticized by those who claim that some difficult decisions may be uncaused, if nothing makes us to choose the way we do, then they are the products of free will. Some people claim that it is possible that unconscious neural events may determine our thoughts and actions. Our decisions could be affected by neural events which are themselves affected by factors such as genes, upbringing, emotional trauma, drugs, alcohol, or brainwashing. The question of whether we have free will is a scientific one which perhaps cannot be answered by our current state of knowledge about the brain.

Many psychologists and neuroscientists think that there is enough scientific evidence to reject free will and claim that our conscious decisions are caused by events that occur before we choose, and which are completely out of our control. They tend to embrace determinism and suggest that our behavior has to follow the known causal laws of nature. Our thoughts and behaviors may be determined in such a complicated way that we cannot describe the underlying processes and adopt the working theory that free will exists.

The argument from psychology is based on the fact that many of our decisions are caused by things that we are completely unaware of, like subliminal messages. But most available studies are concerned with our behavior not with decision-making, especially decisions that we make while we are feeling torn as to which option is the best. They don't show that our decisions are always caused by subliminal factors.

In support of determinism are the studies of Libet, who claimed that brain activity could be detected fractions of a second before his experimental subjects became aware of making a decision. Nevertheless, these studies don't prove that the brain activity is part of the process that causes us to choose in a particular way. To others these experiments only suggest that we don't need to be conscious to have self-generated actions. The studies of Haynes using Magnetic Resonance Imaging (MRI) showed that neural activity often precedes the conscious intention to move. Some experts argue that these studies only show that although sometimes we are driven by unconscious things, we are not always driven by such things. An alternative explanation is that the observed brain activity is an early neural signature of the brain already making a decision. If our actions were determined only by simple interactions we would have a difficult time handling complex deliberations.

Libet studies are based on the knowledge that conscious decisions are associated with a certain type of brain activity called the "readiness potential," the conscious intention to act and the act itself. He had his subjects face a large clock that could measure time in milliseconds and told them to flick their wrists whenever they felt an urge to do so and to note the exact time that they felt the conscious urge to move. What he found is that the "readiness potential" arose about a half a second before the conscious intention to move. The conclusion of these studies is that we don't have free will because there is purely physical, non-conscious brain activity that causes our actions to occur before we make our conscious decision. In reality, though, we don't know that the readiness potential plays a causal role in the production of our actions. We just don't have any idea what its purpose is.

Haynes gave his subjects two buttons, one for the left hand and one for the right hand. He told them to make a decision at some point as to which button to push and the press the given button as soon as they made their decision. He found that there was unconscious neural activity in the brain that predicted whether subjects were about to press either button. Moreover, he found that this activity arose as long as seven to ten seconds before the person made a conscious decision to press a given button.

Compatibilists argue that there is no incompatibility between free will and determinism, free will and determinism can go together. Soft determinism refers to the compatibilist view that holds that determinism is true and we have free will. Many compatibilists claim that free will exists if we are not alienated from our own motivations. If we are not, then we are free regardless of whether these inner states have been determined by previous causes. Other compatibilists argue that we are free if we are able to act according to the good or free of neurotic compulsions.

Chapter 10. Existential Concerns

We didn't ask to be born, we were "thrown" into this world and forced to face several existential issues. Death is one of these issues; it may prevent us from reaching our goals and puts a limit to the duration of contact with friends and family. It is a source of anxiety, and we don't know what happens after we die. Most people try to adjust to these concerns and maintain inner peace within the framework of either the religious or the scientific worldview. Religion helps to deal with the fear of death by suggesting we may enjoy immortality and an afterlife (Schweid, 2006). Nonreligious people don't believe in an afterlife. They place their emphasis on how they live while they are here, and many people (religious or not), believe that having children, creating something useful, or performing good deeds can represent a form of immortality by being remembered or having a positive impact on others.

Another existential concern is the fear that there may be no goal to the universe and no purpose to our lives. There is a human need to understand one's experience and to feel that life has significance and purpose. Religion informs the beliefs of many people about the nature of reality and brings them hope that they may be part of God's hidden plan. Secular individuals have to find value and meaning in contemplating the majesty, complexity and creativity of the universe. Others find meaning in the experience of being connected to everything that exists.

Most people hold that we can act with autonomy and freedom. Many psychologists, however, claim that genetic factors, unconscious processes and cultural influences determine our behavior and that free will is an illusion. Furthermore, they assert that we have to accept that we have a dual nature; we can be aggressive or loving, and exhibit both primitive emotions and the capacity for reasoning. The dark side of our nature may explain why there is so much man-made pain and suffering in the world.

Experiences of disapproval, criticism, rejection and social exclusion can lead to depression and feelings of isolation and insecurity. Many people deal with these existential concerns by activating the attachment instinct, a felt sense that the world is safe, people are caring and the individual is unique and valuable. Attachment helps to ease insecurity and restores the feeling of connectedness to our close relationships. Building up our personal identity and self-coherence helps to fight self-alienation, apathy and boredom.

Thus we face a number of existential concerns which include facing death, the threat of meaninglessness, the need to accept our dual nature, the challenge of freedom, the pain of isolation and insecurity, the search for identity and self-integration and the problems of boredom and self-alienation. We are alone in this world, struggling to tame our dark nature, fighting to survive and reproduce and facing the certainty of death. Existentialists are aware of these issues and remind us that we have no choice but to face them.

Shared worldviews and self-definition help people to defend against these concerns. Some people experience the dreadful feeling of being vulnerable and cut off from the rest of the world and look for safety and security in close relationships. Many people join religious or social networks in their search for meaning and safety. Others, like Whitehead, hold that we are constituted by our relationships, suggesting that we are dependent on the universe for our experiences. This appears to contradict the existentialist's claim that we have to accept the day to day responsibility for creating our own reality. All these problems are part of human existence; it is important to find a constructive and healthy way to deal with them.

Death

Oblivious to the human yearning for permanence, the universe is relentlessly wearing down towards a condition of maximum disorder (the second law of thermodynamics). The fear of death and annihilation and the deprivations that it causes creates great existential anxiety and may be a reason why most people believe in God and why throughout recorded history most religions promote the practice of mourning and the belief in the immortality of the soul. There are individuals that report that they had glimpses of the afterlife after having near death experiences in which they describe feelings of serenity, detachment and the presence of a light. Most scientists, however, believe that these patients are not actually dead; they just have hallucinatory experiences associated with decreased oxygen delivery to the brain. In addition, scientists have found that stimulation of certain brain regions results in detachment from the body or spiritual experiences.

Some scientists are currently working on ways to prolong life span or at least eliminating the ravages of old age. Transhumanists think that one day we may be able to reengineer the human species (Minsky, 2006). By doing so they hope to decrease the pain and suffering associated with aging, lower the cost of health care and allow individuals to pursue new interests or spend more time with friends and family. On the other hand, it could be argued that prolonging the life span is selfish, preventing young people from having their own children. It may also cause stagnation, boredom and a postponement of life decisions. The cost of the treatments to increase life span may be high, and only rich people may be able to afford the high cost of these interventions.

To defend against death anxiety religious persons find comfort and meaning not only in this life but also in the hope that there is an afterlife. They expect that their immortal soul will go to heaven, meeting again friends and family, finding peace and happiness and participating in God's plan. People without religious affiliation find no evidence that the soul exists and that it survives the death of the body and the brain. They find joy and meaning in this life engaging in human relationships, doing creative work, and having the satisfaction of helping others. For them just being alive is worthwhile, they may experience the mystic-like or spiritual feelings associated with being part of a wondrous universe. They claim that some form of immortality is afforded by the continuity of their germinal plasma, the products of their creativity and the memories of friends and family. They recognize that biologically-speaking death has played an indispensable role in the evolution of life; death helps to avoid overpopulation, discard the old and make way for the young.

Many writers have taught us to accept death and reject the irrational means we use to cope with it. For example, Epicurus said that death is nothing to us, since as long as we exist death is not with us; but when death comes we cease to exist. Lucretius asked us to look back at the time before our birth, in this way nature holds before our eyes the mirror of our future after death. Becker claimed that most human activity is designed largely to avoid facing death and to overcome the fear of it by denying in some way that it is the final destiny for man. Heidegger thought that not confronting death makes us live inauthentically.

There are several ethical issues associated with the end of life. Thanatology is the academic study of death in humans, including the circumstances of death, the impact of incurable diseases on patients and their family and the social attitudes toward death. As a result of this new discipline, there have been substantial advances in hospice care and pain management of terminally ill patients. The social and religious attitudes toward death and the rituals involved vary considerably around the world. Death ceremonies often entail central motifs of a culture and their performance usually helps bolster the solidarity and the wellbeing of the group.

There has been considerable debate concerning the definition of death. Initially it was felt that death was the cessation of respiration and cardiac function. After the introduction of cardiopulmonary resuscitation most states accepted a definition of death based on the presence of brain death, that is, when no detectable brain activity is present. More recently, organ transplantation has been carried out utilizing donors after cardiac death. The transition from life to death no longer can be clearly defined

Euthanasia

Euthanasia refers to ending of a person's life actively by using lethal substances or withdrawing life support (active euthanasia), or passively by withholding treatment (passive euthanasia). Passive euthanasia is sometimes acceptable, and some doctors endorse the discontinuation of extraordinary means of life prolongation when death is imminent and it is the decision of the patient and/or the family. Nevertheless, some ethicists argue that passive euthanasia is morally equivalent to active euthanasia. Active euthanasia is illegal but physician-assisted suicide is permitted in a few states and in some countries such as Belgium and the Netherlands. Many people support the law that permits physician-assisted suicide in terminally ill persons who view loss of autonomy as the greatest of all human indignities.

Several arguments have been marshaled to support euthanasia. The most important one is the right of self-determination. We ought to, if we are competent and don't harm anyone, put an end to pain, suffering and economic losses as we see fit. The pain and suffering should not be related to inadequate access to medical care, to the suboptimal application of pain medication or to incorrect diagnosis and therapy. Autonomy implies competence and it is very important that the individual's decision is not based on a mental illness, such as depression, that could be reversed by specific treatment.

Prolonged illnesses may cause financial hardships to the family and to society. There is some fear that economic considerations may result in abuses, "the slippery slope" argument, leading to involuntary euthanasia. Another argument against euthanasia is based on the religious view that it is always wrong to kill an innocent person and that killing destroys something that is the property of God. It is also unclear who should carry out euthanasia: the patient, the family or a physician. Health care workers are trained to help patients and to maintain life and may feel uncomfortable carrying out euthanasia.

Abortion

Abortion is another difficult issue that pits two opposing values, life and freedom, against each other. A key question is when the unborn becomes a "person" with moral rights and legal protection. "Pro-Life" supporters hold that personhood begins at the moment of conception. This Catholic and conservative position assumes that the soul enters the person at the time of formation of a unique genome after fertilization. Others, however, claim that personhood is achieved at different times, such as at the time when the embryo attaches to the wall of the uterus, when spontaneous movements occur, when the heart beat can be detected, when the brain starts to function or when the fetus becomes viable. The last two are hard to define and may run counter to logic in the sense that we don't question the personhood of mentally-retarded people and adults incapable of independent living. The question of when "personhood" begins is a moral question that we must face regardless of whether we believe in any god.

One argument heard from some religions is that abortion and perhaps all birth-control measures interfere with God's design for procreation. This implies that any sexual activity that doesn't have procreation as its intended outcome would be wrong. Thus to avoid all acts that could violate the aforementioned argument, we would have to ban all acts that interfere with human procreation and certainly all sexual reproduction. It is well established, however, that about a third of all pregnancies end in miscarriages; a staggering number of embryos die every year of natural causes and abort spontaneously. It is well established that human mating activity serves an essential purpose beyond mere procreation.

The potentiality thesis holds that the zygote is a potential human being. This is one of the reasons that embryonic stem cell research has been banned or limited in some countries. But it could be argued that the sperm and the egg, and for that matter, any cell in the body, could also represent a potential human being. In addition, we do not defend the right to life of embryos that may be discarded after in -vitro fertilization and those lost during the use of contraceptive devices, all of which have human potential.

The sanctity of life thesis is a powerful argument against abortion, but we seem to apply it selectively. We apply it only to human life; we slaughter many animals every year in order to feed the population or just for the pleasure of hunting. In addition, not all humans are treated equally; the sanctity of life thesis does not apply to individuals executed for capital crimes and those killed during wars. The old religious commandment not to kill seems to be applied only to a certain group of people.

"Pro-Choice" supporters don't see the embryo as having achieved personhood from the moment of conception; they give the rights of the mother priority over those of the embryo. They note that we do not have any objective evidence by which to decide whether an early embryo is a person or that abortion is murder. Most people who find abortion acceptable would limit the conditions under which it could be permitted, for instance when the pregnancy is the result of rape or incest, when the fetus is deformed or when the mother's life is at risk. Other conditions such as lack of resources to raise the child, teenage pregnancy and unfit mothers are also debated.

Animal Rights

Many people object to the killing of animals for sport or to feed the population. Many animals are kept in crowded, unhealthy conditions, are fed the wrong type of food and endure prolonged suffering in the slaughterhouses due to the methods employed in the food industry. Hunting and fishing are considered a sport, not recreational animal abuse, and many animals endure pain and suffering when they are used for medical research. Some religions justify speciesism claiming that all animals don't have a full moral status and that they are expendable and subject to our whims.

Vegetarianism has been suggested by those wishing to end the killing of sentient nonhuman animals and the infliction of suffering upon them. The utilitarian view was advanced by Singer and the moral rights argument by Regan. Singer claims that only a boycott of the meat industry may end the practices used in rearing animals for food.

There are many arguments in favor of vegetarianism. It is well known that vegetarians can live a long and healthy life and that an adult can stay in nitrogen balance ingesting non-animal protein. There are several arguments against meat consumption. Meat represents an acid load which may result in leaching of calcium from the bones and kidney stones. It may also be associated with increased mortality due to premature atherosclerosis because of its high fat content. Finally, the energy cost of producing animal protein is much larger than that used to produce non-animal protein.

The most important objection to vegetarianism concerns the morality of the predator-prey relationship. Because many people don't consider it wrong for wild animals to kill prey, and humans are natural predators, it may be morally acceptable for them to kill animals for food. Other critics of vegetarianism utilize a *reductio ad absurdum* argument; if we have to stop killing animals for meat, then we should try to stop all animal predation. Regan, however, has argued that there is no obligation to prevent predation in the wild because predators are not moral agents.

War

War has brought death and suffering to millions of people including civilians. Clausewitz defined war as an act of violence designed to compel our opponents to fulfill our will. Most wars are initiated to obtain resources or territory, to defend against aggression, for political or social reasons or to protect the rights of minority groups. Social stereotyping and in-group versus out-group behaviors often contribute to violent confrontations between groups.

There are three general philosophical theories for war. The just war theory claims that for war to be justified, it has to be prompted by a just reason; for example, the protection of innocent people or self-defense. Americans have been led to believe that they should support just wars and promote democratic ideals around the world, the "good war" myth. In the twentieth century alone, the United States engaged in over one hundred military conflicts. With the exception of the Second World War, there is little evidence that most of these interventions were morally justified. The realist war theory considers that security reasons and the need to promote self-interest are legitimate reasons to go to war. Their followers do not consider that moral reasons are necessary to justify war and would not rule out pre-emptive strikes in order to achieve their objectives. Pacifists oppose war as means to settle disputes. Principled pacifists find no moral justification for war and pragmatic pacifists hold that the benefits of war never outweighs its costs; however, they allow it in special circumstances.

War is one of the most destructive human activities. We should avoid wars and come together to fight for life not against each other. Any social, moral or political objective achieved militarily should be outweighed by the damage inflicted to the civilian population, the so-called collateral damage. The military uses the doctrine of double effect to justify civilian casualties. They hold that while is wrong to intentionally kill innocent people, it is permissible if it is the unintended side effect of a legitimate military action.

Prison and Capital Punishment

There are millions of people imprisoned around the world. The United States has the highest incarceration rate of any country. Many of these individuals are addicts or mentally ill and are not getting adequate treatment. The question of whether and how punishment for crimes is justified is an important moral issue. It is unclear if ordinary citizens have a right to punish criminals and if this power can be transferred to the government as suggested by Locke.

As mentioned before, there are many reasons to punish criminals. The deterrence model argues that in order to maintain the social order states can deter crime by punishing criminals. It works by instilling fear of punishment in potential criminals and does not require that the offender is actually guilty of a crime. The problem with this argument is that it may neglect individual rights. Because the low rates of conviction for most forms of crime, the state cannot effectively deter crime without committing widespread rights violations. It appears, though, that in spite of the threat of punishment, including the death penalty, the rate of crime in this country has not diminished appreciably. The retribution model holds that the reason for punishment is not to deter crime or maintain the social order but it is because criminals deserve to be punished. It is based on Kantian ethics and the view that individuals should treat others as an end and never merely as a means. Kant's principle of proportionality requires that that the severity of the punishment matches the severity of the crime. Nevertheless, this principle has been criticized because it may lead to morally unacceptable ways to punish criminals, including raping rapists or torturing sadistic individuals. In addition, it has been argued by some individuals that the only legitimate function of government is to protect people's rights not to punish crime.

The mixed model combines elements of deterrence and retribution. According to this model the state can punish criminals without violating their rights because criminals forfeit their rights by violating the rights of others. The principle of proportionality provides a useful guide in determining what rights criminal forfeit. Nevertheless, it is unclear what type of punishment would apply in the case of "victimless crime." Moreover, punishment is not always fair and the rights of people are frequently violated. For example, some innocent individuals are punished and sometimes executed for crimes that they didn't commit. Not only the rights of criminals may be violated but some forms of punishment are disproportionate to the crime committed. Other models suggest that punishment may protect society and helps to rehabilitate criminals.

There has been a consistent trend around the world towards abolition of the death sentence. The United Nations passed a resolution in 2007 calling for a universal ban on capital punishment and the European Union requires that member states do not to adopt the death sentence. At least 3,000 people were sentenced to death in 2007 and 88 percent of the executions took place in five countries: China, Iran, Pakistan, Saudi Arabia and the United States. We rank fifth in the number of individuals receiving the death penalty

The death penalty, after a four-year moratorium, was reinstated by the Supreme Court of the United States in 1976. Many states and the Federal government hold several thousand prisoners under sentence of death, mostly African-Americans. There are a few dozen women awaiting execution. The death penalty was abolished for individuals with mental retardation in 2002 and for juvenile offenders in 2005. In April 2008, the U. S. Supreme Court ruled that lethal injection does not amount to cruel and unusual punishment.

Supporters of the death penalty claim that it deters crime and that it constitutes appropriate punishment. It is also less expensive than keeping offenders in life imprisonment. Those who argue against capital punishment hold that it does not deter crime more than life imprisonment, that it discriminates against minorities and the poor and that it perpetuates the cycle of violence. Wrongful execution is a miscarriage of justice occurring when an innocent person is put to death by capital punishment. In the United States, recently introduced DNA evidence has allowed exoneration of several death row inmates and others sentenced to long prison terms.

Meaning and Purpose

Many people wonder about the meaning of life. Some philosophers claim that it is an important question, whereas others hold that is too vague or that it has no answer. Analytical philosophers think that the question can be partitioned or reformulated into other questions such as: "What is the nature of existence"? "What is the value of life"? "What is the purpose of life"?

The existence of God gives meaning and structure to the life of religious people. Theists believe that life in a godless world is meaningless. Naturalists come in two varieties, pessimistic naturalists affirm that meaning requires the existence of God and optimistic naturalists assert that life can be meaningful even in the absence of God. The latter believe that creative activities, human relationships and the appreciation of the beauty and complexity of nature are enough reasons to give meaning to life.

Some people hold that death helps to give meaning to life, whereas others believe that life is meaningless in the absence of immortality and an afterlife. Naturalists believe that each of us, the species, the planet and even the universe will eventually end, as dictated by the laws of physics. Nevertheless, they think that there is no reason to take meaning away from events, even if they are only temporary, and argue the continuation of the species is an important source of meaning. We are interconnected with the rest of nature, which validates our place in the universe, the atoms in our body will not cease to exist; they will reappear in organic or inorganic form at a future date.

Naturalists hold that just existing and being able to enjoy the beauty and wonders of nature is a gift. Arguably, we are special because we are the most complex and intelligent of all living organisms in this planet, the product of billions of years of cosmic evolution. It is conceivable that we may be the only or the most advanced life form in the universe. In the future we may continue to evolve and increase our knowledge, to harness other forms of energy and perhaps to change the course of cosmic history.

Our minds and bodies exhibit remarkable functions that result from the coordinated interaction of billions of cells and their constituent molecules. We are able to reproduce and transmit our germinal plasma thus becoming a link in the great chain of life on this planet. Our extraordinary and refined senses, especially our vision, permit us to appreciate and enjoy many of the wonders of the universe. Our close relationships are a source of love and support and our good deeds and creativity will be remembered by future generations.

Freedom and Autonomy

Most people believe that we have free will and can exercise our autonomy. Fromm also celebrated human autonomy but understood that some people need to see themselves connected and part of a group and thus "escape from freedom." Many social psychologists, though, portray humanity as weak, defensive and lacking volitional capacities, especially during stressful situations. As previously discussed, many scientists claim that self-caused actions and free will are unscientific concepts and that many of our beliefs, goals and behaviors are only products of unconscious mechanisms.

Insecurity

Personal security is a primary concern for every individual. Experiences of isolation, having to face life's stresses and death alone, are important causes of insecurity. According to attachment theory insecurity in children activates the attachment system. When caring and responsible attachment figures are available children feel safe and connected. Insecurely attached children fear abandonment and seek closeness, avoidant children don't appear to need closeness and fear intimacy and disorganized children exhibit features of both anxious and avoidant children. These existential concerns are also relevant to adults who often develop maladaptive behavioral responses to isolation and social exclusion and feel better when they have close relationships.

Irrationality

We have to accept that we have a dual nature. We exhibit emotions and drives, which are mostly unconscious, and we have the capacity to reason. Emotions originate in the limbic system, whereas rationality localizes to the pre-frontal cerebral cortex. Hume stated that reason is, and ought to be, the slave of emotion, without it there would be no goal-directed behavior. We often don't have control of our emotions, though, and behave in an irrational manner.

We have the illusion that we are in control of our lives and influence the outcome of events by making rational decisions. Many experts, however, have concluded that reason doesn't play an important role in many of our decisions. Thus Haidt claims the reason informs our decisions only in special circumstances, we usually "see" things and then work to find justifications. Margolis and Kahneman distinguish two types of cognitive events that closely resemble intuition and reason. The former is fast, emotional and unconscious and the latter slow, deliberate and conscious. They claim that many of our decisions are made rapidly and intuitively as a result of unconscious events.

Humans recognize that they have a dark side; they are capable of both good and bad behaviors. They may engage in competition as well as in cooperation. In ancient China, Mencius held that humans are benevolent whereas Xunzi claimed that they are basically bad. These conflicting views about human nature were also expressed in the West. Rousseau asserted that human nature is good and Hobbes felt that humans are naturally competitive and violent.

Many philosophers admit that humans can exhibit both good and bad behaviors. Plato claimed that human nature is like the natural union of winged horses and their charioteer. He claimed that while the Gods have two good horses, humans have a mixture: one horse is good and beautiful and the other is neither. Freud proposed a model of the mind which includes the *id* or the innate instincts, the *ego* or the executive function, and the *superego* or the behavioral rules learned from family and society. He also stressed that parts of the mind function at the unconscious level.

The traditional Christian religious explanation for our dark side, based on the teachings of Augustine, is that we are sinners and that the devil turned our moral universe upside down. Satan was responsible for Adam and Eve's disobedience of God and their fall from primordial innocence. The doctrine of incarnation states that God took on a human body in order to suffer a painful death to pay for the sins of humanity.

Selfhood

Although some animals can recognize themselves in a mirror, only humans have a conceptual sense of self. Most of us endeavor to develop a well-defined self. The concept of self, though, is difficult to explain. Freud utilized the same word, "ich," to describe both an intra-psychic structure and the individual's self-experience. His editors, though, translated the word "ich" as "ego." Hartmann defined the ego as the structure that interacts with the id and the superego, and the self as the entity that interacts with objects. Kernberg held that the self is embedded in the ego and is the product of integration of many self-representations. For Jackson the self is identified as, but not the same as, the capacity for an awareness of inner events, a special form of consciousness that appears late in evolution. He called awareness of an event object consciousness, which is linked to subject consciousness or the continuous state or stream of consciousness. Others have claimed that the self is constructed by people around us; it may exist primarily as a conduit to let the social group modify our natural impulses with socially derived and accepted impulses.

On the other hand, Buddhism teaches that we have no-self. It doesn't mean that we don't exist; it means that we don't have an abiding enduring self and we are a mixture of several aggregates. Foucault challenged the notion of the self, thereby dismissing the possibility of biographical study. Some neuroscientists support this view and have embraced the bundle theory after discovering that there is no single center of first-person consciousness in the brain. They believe that we have different brain modules doing different things at different times. Recent functional MRI studies, though, have shown that the medial prefrontal cortex is the center of both our self-concept and the capacity to assimilate the values and beliefs of the social world. Any conflict between the two is resolved by self-control mechanisms.

Childhood abuse, control or neglect, may impair self-integration, especially in those individuals genetically or temperamentally predisposed. The problems associated with lack of self-integration range from mild codependency to severe personality dysfunction. The former occurs in individuals in which their true self was not acknowledged, and compensate by creating a false self. As a result, they lack self-esteem and become other-dependent. They may become overhelpers and perfectionistic in order to be accepted and have serious difficulties in relationships.

Addictions

Many individuals engage in addictions to defend against life's pain, suffering and emptiness. The mood-altering agents or behaviors that they embrace all have in common the capacity to activate the brain reward system. A variety of normal pleasurable stimuli, such as food and sex, constitute rewards that utilize this pathway. They arose during evolution in order to reinforce the need to ingest food, to ensure survival of the individual and to promote reproduction and the preservation of the species.

The main brain's reward system consists of dopamine neurons in the ventral tegmental area and their projections to the nucleus accumbens and the prefrontal cortex, the so called mesolimbic dopamine system. Addictive agents directly or indirectly hijack the underlying reward-related mechanism. The frequent activation of this system may be accompanied in susceptible people by the development of full-blown addiction, which includes tolerance, the negative effect of withdrawal and the craving and anticipatory stage that follows. Some addictions such as drug use and sexual predation may lead to criminal behaviors. Others such as compulsive eating, smoking, exercise and love addiction are tolerated by society.

The most important causes of addictive behaviors are alcohol, smoking, food, drugs and other external sources of gratification used to activate the "reward system" and ward off, at least temporarily, loneliness, anxiety and despair. Unfortunately, these addictions are very costly to the individual and to society and are associated with serious side effects. Some addictions such as smoking and drinking were initially accepted by society. Nonetheless, drinking and driving and smoking in public places are now considered unacceptable.

Alcohol has been used for thousands of years as a mood regulator and a complement to meals and social interactions. Unfortunately, eighteen million adult Americans are alcoholic or abuse alcohol and many more are engaged in drinking patterns that may lead to alcoholism. Abusers do not manifest symptoms of addiction such as craving, dependence and tolerance. There are serious medical consequences of alcoholism including liver cirrhosis, pancreatitis, brain damage, fetal-alcohol syndrome and cancer of the throat, esophagus and larynx. Alcohol is a factor in about half of automobile accidents and acts of violence such as murder and suicide. Each year more than one hundred thousand Americans die of alcohol-related illnesses.

Smoking is encouraged by advertisement, the popular culture and the addictive effects of nicotine. It remains the leading preventable cause of death in the United States. Smoking causes thousands of deaths every year and has a profound economic impact on the health care system. Although the rates of smoking continue to decline, an estimated forty six million Americans still smoked in 2001. Adverse health consequences of smoking include lung and throat cancer, emphysema, pneumonia, atherosclerosis and cardiovascular disease, stroke, low birth weight, cataracts and dental disease. The 2000 Surgeon General's report provides a detailed framework for comprehensive tobacco use prevention and control.

Eating disorders usually affect young women, because of the pressure in society for females to be thin. They may have a devastating effect on the individual and their families. Eating disorders include anorexia nervosa, bulimia nervosa and binge eating. Subjects with anorexia nervosa have low body weight and hormonal irregularities, such as amenorrhea, for more than three months. Bulimic patients may have normal body weight but experience binge eating alternating with purging (vomiting or laxative abuse). These patients may experience medical problems including heart disease, osteoporosis, hormonal and fluid and electrolyte disturbances. The addictive potential of high caloric products, their promotion by advertisers and our sedentary lifestyles may be responsible for the current epidemic of obesity in the developed world.

Substance abuse is a major public health problem. Marijuana, leaves of cannabis plant, cut, dried and packed into cigarettes, is the most commonly abused drug. It is estimated that a third of the adult population has used it. The active ingredient, 9-tetrahydrocannabinol interacts with brain receptors to induce characteristic central nervous system effects. In addition, the drug may cause conjunctival injection, hunger, dry mouth and tachycardia.

Opioids include natural opioids (morphine), semisynthetic opioids (heroin) and synthetic opioids (codeine or meperidine). Cocaine, from the coca plant, is consumed by chewing the leaves, as coca paste or as cocaine hydrochloride powder that can be snorted through the nostrils. "Crack" is cocaine alkaloid extracted from the hydrochloride by sodium bicarbonate and allowed to dry in small rocks. It can be inhaled causing rapid and powerful effects. Before "crack," the alkaloid was extracted with solvents, with the possibility of ignition, and then smoked as the "free base."

Amphetamines are frequently-prescribed stimulants used for the treatment of attention-deficit/ hyperactive disorders and as appetite suppressants. Methamphetamine (speed) is commonly abused. When highly pure (ice) it can be smoked. Its stimulant effects are similar to those of cocaine except that it does not have membrane effects or affects the heart. Phencyclidine (PCP) and ketamine were initially developed as anesthetic agents. They can be taken orally, intravenously or smoked. PCP is known by the street names of hog or angel dust. Hallucinogens include ergot derivatives (LSD) phenylalkylamines (mescaline), indole alkaloids (psilocybin) and 3-4 methylenedioxidemethamphetamine (ecstasy). Some individuals get "high" inhaling aromatic or aliphatic hydrocarbons found in glue, gasoline, and paint thinners. Rohypnol (roofies) and gamma-hydroxy butyrate (GHB) and other sedatives are used for date rape and during "raves." Caffeine addiction may cause sleep disorders and anxiety reactions.

The introduction of national lotteries, large casinos and internet gambling has greatly increased the participation in gambling worldwide. Gambling addiction is diagnosed when the individual cannot control or stop gambling, conceals the extent of involvement, jeopardizes work and relationships and commits illegal acts to finance gambling. Gambling is difficult to resist because it provides excitement, risk-taking and the possibility of getting rich. People gamble even if they know that the odds of winning are weighted against them.

Sex addiction, pedophilia and various forms of sexual dysfunctions are also disturbingly widespread.

Love addiction often originates in families in which the parents don't get along and one of them turns to the child as a source of emotional support, or when the caregiver, because of personal problems, fails to connect emotionally with the child. A parent in an unhappy marriage may commit emotional incest by turning to the child of the opposite sex to obtain the intimacy and companionship that one would normally expect in an adult love relationship. Alternatively, the parent may be unable to make an emotional connection and the child may be left with a chronic fear of abandonment.

The victim of emotional incest often doesn't get the support necessary for the development of the emerging self. His identity is not shaped by its own emotions and sensibilities but by the parent's needs. The emotional incest victim may feel that he is getting special attention but in reality he is being forced to play the role of surrogate spouse, he nearly always suffers from neglect and abandonment because his needs for nurturing and individuation are not met. The left out parent and other children in the family may become resentful and act cool and distant.

As an adult, the emotional incest victim is compulsively attracted to persons similar to the offending parent, an individual who needs to be rescued. This is an unconscious recreation of the parent–child relationship, a way to work out the original childhood situation. In relationships he avoids physical and emotional intimacy and is afraid that he will be drained, engulfed and controlled by the partner. Unconsciously, however, he fears the type of abandonment that he experienced as a child. In order to escape from this situation he may pursue outside interests, including other addictions, or he may leave the relationship and look for other partners.

On the other hand, the individual who was abandoned or neglected by the caregiver feels empty and incomplete because of lack of validation, love and affection. As an adult he fears abandonment and creates a dependent false self that looks for someone to connect with. He may experience very low self-esteem and assigns a lot of time and energy to the relationship, failing to take care of his own affairs. He feels safe in the world only when he clings to another person. Unfortunately, this person is often a victim of emotional incest, a rescuer, avoidant or distancing individual who is afraid of intimacy and never makes an emotional contribution to the relationship. Thus the victim of parental abuse continues to feel abandoned. At an unconscious level, he also fears intimacy because it makes him vulnerable to abandonment. He may cling to his partner to the point of smothering, may become angry and hostile or may surrender and stay in an abusive relationship that often mimics the one he had with his caregiver.

Today's Western society is partly responsible for the problem of love addiction because it tends to sexualize and objectify everything. Love addiction is reinforced in romance novels, movies and love songs and many people are influenced by them. The church also contributes to the problem because of its obsession and repression of sex and the involvement of priests in a large number of cases of child molestation.

Shame

Shame is the feeling of being defective as a human being and not meeting our own personal expectations. In contrast, guilt is about a specific wrongdoing and it arises from a failure to meet the expectations of others. Guilt is a positive emotion in that it helps to recognize our faults and correct our behavior. Bradshaw identifies two types of shame. Toxic shame is a consequence of being self-ruptured and not loving ourselves, whereas nourishing shame is born out of self-love and self-respect.

The victims of emotional trauma often blame themselves for their problems and feel that they, not their parents, are flawed. They are consumed by feelings of inadequacy, of not measuring up to and being worthless and unlovable. The problem is compounded in homes that don't allow the free expression of emotions. Children are often told "don't be a baby" when they are sad, or "you have nothing to be afraid of" when they are scared, their feelings are not validated. Critical remarks from the parents are frequently internalized; they become shaming inner voices that do considerable damage to self-esteem. Hidden feelings impair self-knowledge and self-love and interfere with our ability to love and understand others.

Individuals who feel they are unlovable fear rejection or criticism. As a result, they are afraid of confrontation or expressing feelings that may displease others. They tend to work in the helping professions and are always looking for acceptance. The defense mechanism used by some individuals with low self-esteem is to create narcissistic constructions of perfection, grandiosity, superiority and self-sufficiency.

Anxiety

For us primates, attachment to others means survival, and abandonment spells death. Thus it is not surprising that victims of abuse or neglect, which include a large percent of the population, suffer from fear and anxiety.

Fear results in a series of physiological changes designed to cope with an environmental threat. These changes are mediated by the activation of the hypothalamic-pituitary-adrenal system and the autonomic nervous system. The amygdala is a part of the brain responsible for the processing of fear, it develops before maturation of the cortex; thus children depend on the security and proximity of the caretakers for fear regulation. Anxiety is a more sustained emotion that is diffuse and not directed to a specific entity.

Generalized anxiety, a chronic worry, a feeling that something bad is going to happen, and specific phobias, like fear of flying or certain animals, are very common conditions. Social phobia is triggered by the fear of a negative evaluation by others; stage fright is one of its most common manifestations. Some people suffer from agoraphobia, the fear of situations from which escape is difficult, they may also experience panic attacks. The acute stress disorder usually follows a major traumatic event. Some individuals suffer from post-traumatic stress disorder after an acute event. This condition is characterized by repetitive flash-backs, nightmares and episodes of increased arousal and anxiety.

Some people express anxiety as physical symptoms that are not explained by any known mechanism, their condition is usually diagnosed as a somatoform disorder. Excessive health anxiety, the fear of having cancer or other serious illness is known as hypochondriasis. Some people suffer from pain disorders, including frequent headaches, irritable bowel syndrome and body dysmorphic disorder, a condition characterized by excessive preoccupation with some imagined physical defect or appearance.

Defenses against anxiety include obsessive-compulsive behaviors, dissociation and depersonalization. Obsessive thoughts and compulsive behaviors, like frequent hand washing, may be used to defend and neutralize anxiety. Some psychiatrists, though, feel that obsessive-compulsive disorders are not related to anxiety disorders but represent a separate category of emotional illness. Compulsive behaviors may predispose to addictions, hoarding and busy work. Dissociation is another defense mechanism against anxiety. Dissociation is often described as "spacing out." It may be experienced by some people during life-threatening situations and may lead to posttraumatic stress disorder or to dissociative identity disorders.

Patients with poor self-integration may have experiences that resemble dissociation. Voluntarily-induced meditative or trance practices that are prevalent in many cultures, some of them induced by psychedelic drugs, may mimic and even trigger a dissociative disorder. Some people experience a feeling of numbness and depersonalization, a condition in which the affected individual cannot fully experience reality and feels like an observer.

Dissociative symptoms are intended to distance the self from overwhelming anxiety. Leaving the body or reality may help people avoid experiencing the fear associated with dangerous situations, whereas emotional numbing is a result of neglect and a defense against feeling emotional pain. The common thread of unreality, the loss of self and inability to enjoy pleasurable activities (anhedonia), causes considerable distress to those afflicted.

Boredom and Depression

In our society, entertainers, especially actors, musicians and athletes, are highly valued and extremely well remunerated. It seems unfair that these individuals are paid astronomical figures, several orders of magnitude greater than those paid to more productive members of society such as teachers, firefighters or nurses. The reason why entertainers are overvalued in our society is that their performances provide an escapist, addictive, mood-altering influence on the brain reward system that keeps some people from experiencing the drudgery of daily existence. Entertainment, like substance abuse, helps us to avoid boredom and self-alienation. The latter consists of being separated from one's own essence or nature and forced to live an unfulfilled and meaningless life. Like busy work, entertainment functions as a hedge against ennui or existential boredom, our inability to find meaning in anything.

Movies and musical entertainment may appeal to our basic sexual and/or aggressive instincts. The basic unit to measure time in music is called the beat, the regular pulsation that puts us in touch with our body rhythms and our ancestral past. Many competitive sports may appeal to our violent nature. Some of them, such as football and boxing, are so violent that they may cause brain injury or even death, yet they are still permitted and promoted by society. They are not too different from the gladiator events and animal fights staged in pre-modern times to amuse the populace. Other sports, like car racing, are not only dangerous but also wasteful of needed energy resources.

During entertainment events, when the viewer is already highly stimulated, strategically-placed advertisements provide an economic benefit to the sponsoring corporations. Advertisements stimulate the unbridled consumption that drives the economy. Large corporations use the latest marketing techniques to promote their products and maximize their profits. They take advantage of the "herd instinct," the mindset associated with being part of a group larger than ourselves and letting the behavior of the group do the thinking for us. These corporations often disregard the health and safety of the public in order to promote potentially harmful products, such as tobacco, alcohol and junk food. They threaten freedom by controlling network news and changing their scope. These newscasts are designed to block substantive information and analysis of current issues. Corporations also have an inordinate influence on the political process through monetary contributions to candidates, political action committees and lobbyists.

Self-Deception

Deception is common in nature and many animals use it during courtship and when protecting resources and territory (Trivers, 2011). Deception, including self-deception, is also common in humans. We deny thinking about our mortality, the dangers of risky behaviors, climate change and other stressful information in order to avoid the anxiety they cause. Self-deception is so prevalent perhaps because it protects happiness and self-esteem and confers the evolutionary advantage of helping to deceive others. Thus, most people rate themselves to be more attractive, moral and competent than others. When feeling empowered, individuals tend to neglect the views of other people and claim the high moral ground.

Lowering the status of out-groups helps to degrade their image and may justify aggression towards them. This behavior starts in childhood when we learn to divide others on the basis of factors such as gender, race, ethnicity, status or attractiveness. Throughout life we continue to create false personal and historical narratives. Because history tends to be written by the winners, historical accounts of past conflicts tend to reflect the views of those who prevail.

Other forms of self-deception are imposed from the outside. They include the creation of false memories, hypnotic trance and the placebo effect. When people who witness an event are later exposed to new and misleading information about it their recollections often become distorted. False childhood memories created by overzealous mental health workers have been responsible for many reports of alleged sexual and physical abuse. Hypnosis and the placebo effect are examples of self-deception that result a direct health benefit. The placebo effect is consistent with cognitive dissonance: the more we commit to a position, the more we need to rationalize our commitment to it, and this rationalization leads to a greater effect.

Prejudice

The origin of prejudice may be traced to the self-preservation instincts that evolved in our ancestors. The ability to stereotype a predator would have been expected to increase our ancestors' chances of survival. A specific type of social cognition involves the inference of another's mind, commonly referred to as Theory of Mind. We compare ourselves to others in order to identify with our peer group, to gain self-esteem and to know where we stand.

Many people believe that humans can be separated into biologically different races. Moreover, some people have used these differences to claim that some races are superior or more intelligent than others. These arguments are often used to support racism, ethnocentrism and the interests of the dominant culture. Social scientists, however, claim that race may be a socially constructed taxonomy that doesn't merit consideration. They refer to ethnicity as self-identity with a group defined in part by racial admixture, geography, culture, religion and language.

It is clear that all humans belong to only one species. Studies of mitochondrial DNA and the Y chromosome have proven conclusively that most modern humans can be traced to a migration of a small group of anatomically modern individuals who first appeared in Africa around two hundred thousand years ago. One group of migrants stayed close to the coast settling in the Indian subcontinent and from there moving on to Australia and Indonesia. Another group established themselves in central Asia and subsequently migrated to Europe and to Siberia. Less than fifteen thousand years ago, during the last ice age, members of the latter group crossed Beringia, the frozen land uniting Asia with the Americas, and from there they populated the rest of the continent. They are the ancestors of all Native Americans.

It appears that different human populations were separated long enough to allow some genetic diversity to take place. Many scientists think that the reason for some of the differences among populations is the influence of climate and geography. As populations migrated northward, their dark skin, which had been helpful to shield them from the sun's damaging ultraviolet radiation, now prevented the production of vitamin D which resulted in rickets. Individuals with lighter skin were more likely to survive and reproduce. In cold climates people developed shorter limbs and a heavy trunk in order to conserve heat, whereas in hot climates people tended to become tall and slender. Certain facial features such as slanted eyes could have evolved to protect the eyes from the sun's glare reflected from snow; long noses in people living in arid lands may have developed in order to better humidify the inspired air.

We would live in a better world if we could modify human nature to take advantage of our superior cognitive abilities to minimize violence, inequality and prejudice. Our basic needs could be satisfied with simpler and healthier lifestyles which include spending more time with friends, family and community, achieving a more egalitarian society and developing a deep respect for the sanctity of nature. We could take measures to limit population growth and protect the environment. We could also tax and reduce the amount of junk food, educate the public about the adverse health consequences of a poor diet, reduce the vicarious sexual practices and learn to share our wealth.

Humans are the most violent animals on the face of the Earth and it will be very difficult to curb violence and warfare. We could, however, reduce the impact of militarism, propaganda and fear-mongering, curtail the influence of alpha males in positions of power and recognize the humanity of our enemies and minorities. It will be important to develop international organizations to foster economic growth in developing countries, to eliminate gross inequalities in income and wealth, and to enforce disarmament and conflict resolution around the world.

Randomness

The ways we analyze and decide about situations involving chance pose a difficult challenge to most of us. We tend to create patterns about events that are controlled by chance and make fast and intuitive decisions that are not supported by reason. Our imagination often fills gaps in our visual and non-visual perception. The typical patterns of randomness are routinely misinterpreted and whenever the basic principles of randomness are applied they lead to conclusions that prove counterintuitive.

Probability and statistics were designed to deal with randomness. The basic principles of probability began with the work of Cardano in sixteenth century Italy who developed them to deal with games of chance. He described the importance of evaluating the sample space or the set of all possible outcomes. For example, when we toss a coin the sample space is two (heads or tails) and the probability of getting heads is one in two. When we toss two coins, though, the situation becomes more complex. The number of heads that turn up in those tosses can be zero, one or two; and since there are three outcomes, the chances of each should be one in three. Nevertheless, when we examine the sample space, or all possible outcomes, we notice that there are actually four outcomes (head/head, head/tails, tails/head and tails/tails). Thus the probability is one in four for zero or two heads and one in two for one head.

The way results reflect underlying probabilities when we make a large number of observations, the law of large numbers, was described by Bernoulli. If we throw a coin ten times, we may get six or seven heads in a row, but if we throw it a thousand times the probability of getting heads would be very close to fifty percent. The gambler's fallacy refers to the false impression that an event is more likely to occur because it has not happened recently. Thus many people don't understand that if we get seven heads in a row the probability of getting tails in the next toss is not increased; it is still fifty percent.

Bayes's theory helps us to infer underlying probabilities from observations and then adjust them on the basis of new data. Thus the theory uses new information to prune the sample space. It is useful for instance in the evaluation of the results of medical tests when the knowledge of false positives and the prevalence of disease are known. It also explains the prosecutor's fallacy: the probability that x will occur if y occurs will generally differ from the probability that y will occur if x occurs. Thus the probability that a perpetrator of domestic abuse will go on to kill the victim is small, but among battered women murdered, most of them are killed by the abuser.

Probability is concerned about predictions based on fixed probabilities, whereas statistics is concerned with the inference of probabilities on the basis of observed data. The latter was developed by Laplace and others to deal with the imprecision in measurements introduced by random errors. For instance, failure to take into account the error of replicate determinations when analyzing the results of electoral polls or unemployment figures often leads to the wrong conclusion. The central limit theorem explains why the normal distribution is the correct error law. The standard deviation around the mean shows how close to the mean a set of data clusters. A standard error of five percent in a poll indicates that if we repeat the poll, ninety five percent of the time the result would be within five percent of the original value, however, there is a five percent chance that it will fall outside.

What appears to be a specific pattern in events, such as market trends, athlete's performance and disease clusters, can happen as a result of sheer chance. Scientists get around the problem of identifying false patterns in nature by developing methods of statistical analysis to decide if their observations support a particular hypothesis. One of the most commonly used techniques, significance testing, was introduced by Fisher in the early part of the twentieth century. The p value, though, measures whether the observed result is due to chance but doesn't answer the question of what are the odds that the hypothesis is correct. Scientists often don't consider the importance of effect size and fail to utilize the Bayes theorem to infer plausibility of outcome.

Non-scientists tend to use short cuts and intuition to discern specific patterns and sometimes arrive at the wrong conclusion. Our perception that certain patterns are real, not the product of chance, is probably due to our need to avoid randomness and to have some control over our lives. Instead of trying to prove that a perceived pattern is due to chance we often look for data that would confirm our illusion, the so-called confirmation bias. Hindsight is usually easy but connecting the dots before an event happens is much more difficult because of the complexity of available information and the possibility that random events may take place.

CHAPTER 11. RELIGION AND ITS SHORTCOMINGS

Religious beliefs are so prevalent that many scientists have postulated the existence of a God gene. In a 2012 survey by the Pew Research Center's Forum on Religion & Public Life, eighty-four percent of the world population was found to be affiliated with a particular religion. Thirty-one percent were Christians and twenty-three percent Muslim. In the following we pay more attention to Christianity because it has the most adherents and is familiar to most people in the West. We must acknowledge the appeal of religion to a large segment of the population. Nonetheless, most of the population once believed that the Earth was flat or that it was the center of the universe, beliefs that have been proven false.

In the aforementioned survey, the unaffiliated group came in third place; their number was close to that of Catholics. Most of their members lived in the Asian–Pacific region, Europe or the United States. Only six percent of individuals who claimed no religious affiliation were atheists or agnostics; the rest were spiritualists, skeptics, free thinkers or humanists. In the United States, though, the number of unaffiliated people continues to grow at a rapid pace, especially among the young and well educated; in the last survey by the Pew Research Center, thirty percent of Millennials reported no religious affiliation.

People who are religious posit the existence of God or a supernatural agent to explain the origin and meaning of life and the mysteries of the universe. Their beliefs, however, are often based on faith or revelation, not on evidence or rational proofs. Philosophy is an exercise in reason and inquiry; it cannot prove any truth about the world, only which beliefs are more rational. On the other hand, science can give us approximate models of the world with the information that we get from our senses and other tools. These models can be confirmed or rejected by empirical means. Scientific theories, however, cannot be proven to be true; they can only be proven to be false. In addition, it is difficult for science to explain certain human experiences like romantic love or the color of a flower. Nonetheless, science is the best description of a reality that we possess; it is responsible for most of the advances of civilization in the last four hundred years.

Because nobody knows the absolute truth about the nature of reality, and our beliefs are influenced by the prevailing culture and by psychological factors such as cognitive biases, self-deception and prejudices, it is not surprising that there are so many religions. There are thousands of religions; however, not all of them could be right. Pluralists believe that all religions have the same focus: they are simply different responses to one divine reality. Religions often contradict each other and even within one religion there are subgroups that are incompatible with each other. For example, Christianity is divided into three major groups: Catholic, Protestant and Eastern Orthodox, and they themselves are divided into hundreds, perhaps thousands, of branches. Their tenets may influence how people think about science, morality, political philosophy, economics and sociology.

Our minds may work in a way that makes us susceptible to supernatural beliefs (Shermer, 2011). Primitive man created supernatural beings as the agents at work in nature. An active agent detection device may have helped to create these beliefs. Organized religion began when primitive beliefs were codified and given the structure of a God-given plan. The division of labor in early agricultural societies permitted the emergence of priests and religious leaders.

The cohesion of religious groups had the evolutionary advantage of protecting members from outside groups and decreasing existential stress. The latter may have a survival value in that it protects the immune and cardiovascular systems from the deleterious effects of stress hormones. Many naturalists hold that religious belief in general may be an evolutionary product.

Religious beliefs also have a consolatory aspect. These beliefs help us face mortality and make sense of the powerful and mysterious forces of nature. Our ancestors created parent-like figures that would protect them against life's stresses and perils and promise an afterlife. Hume claimed that the primary function of religion is to console us in our travail. Marx argued that religious ideas are a product of economic and social factors. According to Freud, our religious beliefs may be rooted in the infant's helplessness, the need for attachment and the longing for a protective father. Theory of Mind, the idea that other people have minds just like our own, could explain why these beliefs spread throughout the population.

There are many other reasons for the high prevalence of religion including the indoctrination of children, the social and political pressures to conform, the value of traditions and rituals and the lack of scientific literacy. Religions produce myths to live by and rituals to give structure to life. Most people automatically adopt the religion of the country in which they were born; their decision-making is not really free or rational. Religion is supported by conservatives because they are afraid of any changes that they feel could threaten their safety. The most important reason, though, is that religion gives people an easy way to allay their fears in a world they perceive as hostile and indifferent; it enables them to interpret their experience of the world and to account for its meaning. Believing in God provides answers that many people find adequate to satisfy our most important existential concerns which include finitude, meaning, identity, loneliness and autonomy.

Religion is so central to life's purposes that Voltaire (considered to have been a deist) said that if God didn't exist, we would have to invent one. The existence of God offers a relatively simple explanation of how the universe originated, how life arose spontaneously from inanimate matter and how it is that non-physical entities like the soul exists. For religious individuals, God is also the origin of the incredible variety and complexity of life forms that exist and the reason why the universe appears to be fine-tuned for the creation of life. Believers don't have to elucidate how a few grams of brain tissue create the experiences that we have: they are manifestations of the soul. They don't need to wonder how to act; they are expected to be responsible for their actions and to follow God-given moral rules. They don't have to despair when facing the suffering and injustice that exists in the world, God will protect them and listen to their prayers. They don't have to worry about death: souls are immortal and God will reward them for their good deeds in the afterlife.

Arguments against the tenets of religion are shown in Table 1. It is difficult to comprehend the existence of an entity that is infinite, eternal and immaterial. God's mode of being may be so foreign to us that usage of the word "existence" may be misleading. A lot of ink has been spilled over issues that are impossible to prove or disprove and some of the conclusions that have been reached may be fundamentally flawed or incoherent. Nonetheless, many religious claims have been made and skeptics have raised many arguments against them. These arguments include the inadequacy of the "proofs" of God's existence, the alternative explanations of the fine tuning argument, Euthyphro's moral dilemma, the religious pluralism thesis, Hume's view on miracles, the evidentialists' response to faith, the doubts about the origin and veracity of scripture and the conflicting views about theistic claims, including the hiddenness of God and the problem of evil.

TABLE 1. ARGUMENTS AGAINST SOME OF THE TENETS OF RELIGION

The inadequacy of the "proofs" of God's existence
Scientific explanations for the fine tuning argument
Euthyphro's dilemma
Religious pluralism
Hume's views on miracles
Evidentialism versus faith
Doubts about Scripture
Conflicting views about theistic claims
The hiddenness of God
The problem of evil

The "proofs" about God's existence were spun at a time in which we had very little knowledge about the nature of the world. As previously discussed, the premises of the first cause argument, also known as the cosmological argument, have been challenged. The argument states that absolutely everything has been caused by something else prior to it; nothing has just sprung into existence without a cause. We assume that the universe is the result of a long series of cause and effect that led to its being as it is. If we follow this sequence back, we find an original cause, the very first cause. This first cause, the argument tells us, is God. This argument contradicts itself; it argues that there can be no uncaused cause, and that there is one uncaused cause: God. Its proponents have failed to prove that there is a self-existing being; we don't know that the principle of sufficient reason is true. Hume's famous account of causation severs any necessary link between cause and effect and undermines Aquinas supporting argument. The first cause argument also assumes without evidence that the effects and causes could not possibly go back forever in what is called an infinite regress.

The design or teleological argument states that everything in the world bears evidence of having been designed for a purpose. It is known as an argument from analogy because it is based on the similarity between two things, it relies on the idea that a designed object like a watch is in some way similar to a natural object like the eye. This argument has been undermined by the theory of evolution and by the weaknesses inherent to all arguments from analogy. The argument doesn't prove theistic claims; it doesn't support monotheism, or the view that the designer was all good and all powerful.

The ontological argument doesn't rely on evidence, it claims that the existence of God necessarily follows from the definition of God as a supreme being; a perfect being must include existence as one of its properties. St. Anselm defined God as "that being than which nothing greater can be conceived." The argument's most familiar version was given by Descartes who claimed that the concept of the non-existing, supremely perfect being is a contradiction. Because we can conceive of this being, it follows from the very definition of it that it exists. Kant responded that existence is not a property of anything, but a condition of anything having a property. A recent version of the ontological argument was given by Plantinga. He does not claim that it proves that a god exists, but establishes that it is rational to think that one exists. Subsequently, he has argued that the belief and existence of a deity is "a basic belief," a capacity that everybody has to sense the divine. The most important criticism of the ontological argument is that it seems to allow us to define all kinds of things into existence. We cannot define our way to truth.

Some thinkers have defended a variant of the teleological argument known as the fine tuning argument. This is the view that the chance of the world to have the physical parameters to permit life is so small that we can conclude that the world is the work of a supreme architect. Nevertheless, anything which is statistically unlikely still can happen, the laws of physics that appear to be fine-tuned for the development of life may have arisen naturally. This principle has been rejected by many physicists who claim that the constants of nature apply only to our particular section of the multiverse or that the universal constants may be changing or are interconnected. Some scientists have argued that most of the universe is inhospitable to life and that life forms have adapted to changes in our planet in order to survive, rather than the other way around.

The claims that God is the only source of moral rules and that we have free will and are therefore responsible for our actions have been rejected by many philosophers and scientists. Euthyphro's dilemma for those who believe that morality is derived from God commands is as follows: Does God command what his preference is because it is morally good? Or does God commanding it makes it morally good? There are many good people who don't believe in any God; morality more likely is the result of innate dispositions supplemented by consequentialist societal rules. Christian ethics is based on the assumption that we have free will, however, some experts hold that free will may not exist and that our actions are caused by factors that we are unaware of.

Some religions are exclusivist and claim that only they have the truth. Pluralism is the view according to which one religion is not the only or exclusive source of truth. Some people also believe that God is the main source of meaning and purpose in the universe. The existence of a deity bestows significance or purpose to our existence; there is no evidence, though, that the lack of religious beliefs may compromise the possibility of finding meaning. For secular humanists the meaning of life is not related to supernatural influences. Humanists affirm our ability and responsibility to lead an ethical life of personal fulfillment that aspires to the greater good of humanity. For humanists, meaning is to be found in the beauty and wonders of the universe, relationships and work, even if they are provisional and contingent.

Many individuals claim that religious experiences or miracles support the existence of God (Kelly, 2004). While we cannot do scientific tests on "what people believe they experienced," both logic and scientific experiments suggest such claims are ill-founded. Hume asserted that it is not rational to believe in miracles, violations of natural laws by God, because physical laws are based on observational evidence which is more reliable than personal testimonials. The effects of intercessory prayer have been scientifically evaluated in a few studies, and the most rigorous studies to date suggest that prayer has no beneficial effect on health outcomes. If faith healing really worked, we would have already seen its widespread application in modern health care practices. The claims of out-of-body experiences or visits to the afterlife have been debunked by experts who have offered alternative explanations based on neurological science.

Some people believe that faith alone can provide justification for religious beliefs, even if there is no reason or evidence to support them. The definition of faith varies considerably. Russell defined it as a conviction that cannot be shaken by outside evidence. The Bible asserts that faith is evidence of things not seen. Pascal's Wager holds that we should believe because it maximizes our self-interest. The objections to Pascal's Wager are that we cannot decide what to believe, that self-interest may not be the right kind of belief, and that it amounts to wishful thinking. In addition, it applies only to Christians and doesn't include other religions. Evidentialists claim that it is wrong to believe in anything upon insufficient evidence. It may be wrong because it goes against the ethics of belief and makes us credulous and behave like primitive men.

Doubts about the Veracity of Scripture

An important source of support for religious beliefs is found in the scriptures and sacred texts. These texts include the Torah for Jews, the New Testament for Christians and the Quran for Muslims. None of these books offers sound evidence for the existence of God; these texts reflect the views and needs of ancient believers that lived in pre-modern societies. The texts reflect the writings of many individuals over long periods of time. They are not reliable and their narratives are incompatible with modern scientific knowledge. Scholars have found that some of the scriptures are internally inconsistent and historically inaccurate. Although conservative believers continue to defend the literal reading of Scripture, some have adopted a more liberal and symbolic interpretation.

About a century after the death of Jesus, the Hebrew Bible was more or less cast in its present form. It consisted of twenty-four books. The Law or the Torah refers to the first five books (the Pentateuch): Genesis, Exodus, Leviticus, Numbers and Deuteronomy. These five books are at the beginning of what Christians have long referred to as the Old Testament.

The Old Testament (Hebrew Bible) begins with a description of the myth of creation, the story of Adam and Eve and their expulsion from the Garden of Eden. Old Testament stories like the Flood, the Tower of Babel and the parting of the Dead Sea were borrowed from other cultures. Many of the prophecies mentioned in the Old Testament never took place. The stories of Abraham and Moses and the legends of Saul, David and Solomon were rewritten in order to strengthen the authority of religion after the return of Jews from the Babylonian exile. No definitive archaeological evidence, though, has been found of an Israelite presence in Egypt at the time of the Exodus or of the large empires of Kings David and Solomon.

The religion of the Jews was monotheistic; it affirmed that they were the Chosen People and that they had a special covenant with God. In turn, they were supposed to obey the Law and to worship God by making sacrifices in the Temple in Jerusalem and by prayer and study in the synagogues. Judaism included several groups like the Pharisees, Sadducees and Essenes. Unfortunately, their land was frequently occupied by foreigners; their kingdom was surrounded and controlled by powerful and oppressive empires. Several groups of Jews advocated the forceful overthrow of these invading empires; they conducted several unsuccessful rebellions against the Greek and Roman occupations. The latter broke out in 66 CE and ended with the destruction of the Temple and a massive slaughter of Jews. Jesus himself probably practiced a form of nationalistic, apocalyptic Judaism prevalent at the time; he was crucified by the Romans for treason and sedition.

The New Testament contains twenty-seven books, written in Greek, by fifteen or sixteen different authors, between the years 50 and 120 CE. The first four books are the Gospels which contain stories about the life and death of Jesus. The next books in the New Testament are the Acts of the Apostles; these portray the spread of Christianity throughout the Roman Empire. The next section contains letters written by Christian leaders to various communities and individuals, discussing beliefs, practices and ethics.

The New Testament concludes with the Book of Revelation which describes a series of events leading up to the destruction of this world.

There are no reliable sources indicating that a historical figure existed whom we can equate with Jesus; he left no writings. It is unclear who decided to include only the four gospels of the New Testament and left out other important writings. The authors remained anonymous; their stories are based in oral traditions that were written down many years after the death of Jesus. The writers didn't claim to be witnesses of the events they narrate. The gospels differ in content and sometimes contradict each other. It appears that the New Testament emerged out of the conflicts among early Christian groups and the dominance of proto-orthodox position over that of the other groups.

The main person responsible for the creation of Christianity, a religion about Jesus, was the apostle Paul, a well-educated Jewish merchant who lived abroad and travelled extensively through Asia Minor. He wrote in the Greek language and preached in many parts of the region. He taught that the death and resurrection of Jesus healed the original sin, giving both Jews and gentiles the opportunity for salvation. Initially, Christianity was a Jewish sect, but by the end of the fourth century it had separated completely from Judaism. After the conversion of the Emperor Constantine in 312 CE, Christianity spread rapidly and eventually became the official religion of the Roman Empire.

During the early part of the fourth century, only about five percent of the population of the Roman Empire was Christian. The conversion of Emperor Constantine and the proclamation of Christianity as the official religion of the Roman Empire by his successor was a defining moment in the history of Western civilization. Christianity brought a spiritual message of hope and salvation to the poor and oppressed populace and was a religion available to everybody. The growth of Christianity throughout the Mediterranean world was so rapid that at the end of the fourth century, half of the population of the Empire was Christian. Constantine also convened the Council of Nicea, which established the orthodox Christian creed. The belief system that emerged from this Council required many compromises and the adoption of paradoxical doctrines, such as the Trinity. By that time the Church was less diverse, as most heretic groups like the Ebionites, the Marcionites and the Gnostics had been suppressed. Roman administration in the west dissolved in the face of increasing barbarian invasions, but in the east, Byzantine civilization, combining Greek and Roman cultures, flourished. The separation between eastern and western parts of the empire became permanent after the death of Emperor Theodosius I in 395 CE.

The Constantinople patriarchs governed the eastern portion of the empire under the protection of their emperor and distrusted the growing power of the western popes. The schism between Rome and Constantinople that took place in 1054 was an important symbol of the gradual separation of the two churches. The Orthodox Church centered in Byzantium eventually came under the authority of the Islamic rulers when Constantinople (now Istanbul) fell to the Turks in 1453.

The Catholic religion dominated Western Europe for about one thousand years, but in the 16th century an ideological crisis took place leading to the Reformation. The Reformation was initiated by Luther who attacked the priestly authority and the sale of indulgences. Protestants, as they were called, were encouraged to read the Bible and to interpret it for themselves. For them, salvation depended on a personal relationship with God. The second phase of the Reformation was initiated by Calvin who completed the break from the Catholic Church by rejecting the authority of the clerics. This was followed by the Counter-Reformation; the Council of Trent issued a series of decrees to end corruption in the Church and to reassert the Catholic dogma.

The conflict between Protestants and Catholics led to many wars and acts of violence that have continued until the recent times. Most of the philosophers of the Enlightenment were religious, but in the works of Spinoza, Hume and Kant it is possible to discern that religion was being subjected to rational inquiry. By the beginning of the twentieth century, the sentiment against Christianity started to surface as illustrated by the works of Feuerbach, Marx, Darwin, Nietzsche and Freud.

The third large monotheistic religion originated in the Arabian peninsula during the seventh century. Muhammad, a merchant from Mecca, claimed that God (Allah) spoke to him through an angel. His followers began to write the received messages and compiled a book that became known as the Quran. This book was the main scripture of a new sect that eventually became known as Islam (meaning "surrender"); a Muslim was a man or a woman who had made this submission to Allah. Islam maintained some Judeo-Christian traditions and demanded that human beings behave to one another with justice, equity and compassion. Their attitude was expressed in the prostrations of the ritual prayer that Muslims were required to make several times a day. Muslims were required to give a regular portion of their income to the poor in alms. They were required to fast during Ramadan, to remind themselves of the privations of the poor and to travel to Mecca once in their lifetime for the Hajj pilgrimage. Judaism and Christianity also specify versions of fasting and penitence, prayer, charity and alms-giving, and similar practices.

The new religion united the different tribes and eliminated class differences. Muhammad's successors Abu Bakr and Umar ibn al-Khattab invaded and conquered the Byzantine and Persian Empires. Within ten years of the death of Muhammad, his followers had created a large empire that extended from the Pyrenees to the Himalayas. They respected other religions and referred to Jews and Christians as "People of the Book."

The ensuing struggles for power, though, led to many political assassinations including that of the third caliph Uthman which triggered a five year civil war. Muhammad's son-in-law, Ali, was selected as the new caliph. Ali's rule was not accepted in Syria and the opposition to Ali was led by Muawiyyah from his capital in Damascus. His army eventually eliminated the opposition; in 661, Ali was murdered by one of his enemies. This event was the beginning of an enduring rift between the Shiites, those who remained loyal to Ali, and Sunni Muslims.

The Umayyad dynasty, which started after the death of Ali, was responsible for the transfer of the capital of the empire from Medina to Damascus. In 733 the Abbasid faction began to muster support in Iran and in 749 they conquered Kufah and overthrew the Umayyads. Baghdad became the center of the Caliphate until the tenth century when the empire began to fragment and many independent rulers emerged. Subsequently, the Seljuk Turks seized power and came to a special arrangement with the Caliph. The Christian Crusaders conquered Jerusalem in 1009 and maintained a foothold in the Near East until the 13th century. Beginning in 1220 the Mongols invaded the region from the east and caused great devastation. Many of the invaders converted to Islam. However, the Moorish Kingdom of Granada was lost to the Christians in 1492, ending the Muslim rule in Spain.

Three major Islamic empires were created in the late fifteenth and sixteenth centuries: the Safavid Empire of Iran, the Moghul Empire in India and the Ottoman Empire in Anatolia and nearby regions. The Ottomans conquered Constantinople in 1453, reached their apogee under Suleiman the Magnificent in the mid-1500s, and continued until the First World War. Then their empire was divided between the French and the British.

There are conflicting views about the tenets of theism, and some of them may be internally inconsistent or self-contradictory. According to Swinburne, God is a person without a body, present everywhere, the creator and sustainer of the universe, a free agent, omnipotent, omniscient, all good, a source of morality, immutable, eternal, necessary, holy and worthy of worship. The best we can do is to analyze these attributes analogically. God is a person, but one who is only spirit. To call God a person suggests that he possess something analogous to the personal characteristics that we associate with humans. The problem is that certain of God's attributes, like necessity and eternal, are not analogous to human ones. Theists also believe that God takes special interest in them and will listen to their prayers. God is all-knowing, all-good and al-powerful, he can do anything that is logically possible. He is beyond time but the created order is temporal. God created the world and sustains it by setting in motion natural laws or through miraculous interventions. He is the source of moral order, thus infusing value and meaning into the creation.

Many people have pointed out that the concept of the theistic God outlined above is internally inconsistent or self-contradictory (the impossibility argument). God cannot be all these things, even if we exclude things that are logically impossible. For example, God cannot perform an act which entails his own limitation such as creating a rock too heavy for him to pick up. Another way of arguing against the existence of God on the grounds of impossibility consists in showing that that a particular attribute just cannot be the case. The meaninglessness argument states that we have no idea what we intend when we use the word God. The hiddenness argument means that God doesn't reveal himself to humanity.

Many people have a sense of despair when God doesn't answer their prayers. Most people have never experienced a revelation of God. If God exists, he should reveal himself to humans so that we know that he is real. The explanation of his hiddenness is either that God cannot or will not reveal himself, or that he doesn't exist. If God cannot reveal himself, his omnipotence may be called into question. And, if he would not, then his benevolence must be doubted. The contemporary theistic theologian Swinburne responds to these arguments by saying that God chooses to be hidden because a relationship on his part would be unduly coercive and would reduce humans to moral slaves. However, this line of reasoning implies that the moral standing of an atheist is more pure than that of a theist.

Another theistic objection to the atheist hiddenness argument is that God does reveal himself, and many people in all religions have claimed to have a personal experience of him. Atheists retort that subjective experience may not be trustworthy. Another theistic justification for the hiddenness of God is that he deliberately refrains to reveal himself out of consideration to us. Full revelation would either damage our cognitive abilities or overwhelm our comprehension. As described below, the most important argument against theism is the observation that there is so much natural evil, the existence of so much unnecessary pain and suffering, in the world.

The aforementioned arguments are directed to the features of the God of traditional theism, the personal deity worshiped by the Judeo-Christian religions. Just as the standard theistic arguments don't establish the existence of a theistic deity, the shortcomings examined above don't disprove the existence of God. Some contemporary theologians argue that God exists but doesn't have all the attributes mentioned above. It is possible that Deism is true and God created the world but he no longer interacts with his creation. God may also be all there is, is in the process of becoming or existed at one time but no longer exists.

The Problem of Evil

The problem of evil is the most important argument that has been raised against theism. Evil refers to anything that inflicts severe physical or mental damage which is undeserved and serves no salutary purpose. It can be a natural force (natural evil) or the actions of another human being (moral evil). Theists alternate between angry rebellion against what appears to be the indifference of the deity to suffering, and submitting in silence to an all-powerful God. The uneasiness and even despair a theist can experience from the existence of evil was captured by Dostoyevsky's character Ivan Karamazov. Although Dostoyevsky himself was an intensely religious man, Ivan, the atheist, truly speaks for him when he complains in anger and misery about the suffering in life and the indifference of God. One of the earliest formulations of the problem of evil came from the Stoic philosopher Epicurus. He claimed that either God is less than perfect, even malevolent, or that he doesn't exist. Many thoughtful people have been concerned about what appears to be God's indifference to natural or moral evil. Darwin became even more acutely aware of the inherent cruelty of nature as he observed the ichneumon wasp. The female of their species lays her eggs inside living caterpillars which she has paralyzed with a sting. The larvae hatch inside the live caterpillar and devour it from the inside out.

Theodicies are attempts to explain why a good God permits evil to exist (Loftus, 2008). Some religions dismiss the problems of evil by holding that the claim of evil is illusory or unreal. The obvious objection to this denial of the reality of evil is that it flies in the face of ordinary experience. No reasonable person would deny that torturing or murdering children or systematic slaughter or genocide is immoral and wrong.

Rather than denying that humans do commit immoral acts, some theists attempt to distance God from this. The most popular way of doing so is by invoking human free will, by claiming that God gave us the faculty of free will although he was aware that this gift could be used to make wrong choices. The argument would be that God believed that the goodness of free will outweighs the potential of evil that it could allow. Nevertheless, the existence of free will is unproven; and if it does exist, the idea that it came from God of course cannot be proven; and even the idea of "evil" is problematic. What is evil, what is criminal, what is mental illness, and what is an act resulting from naïve belief in what a respected person tells us will lead to a greater good — such as going to war?

Can we posit that divine interference with human free will is a greater evil than blocking free will? Should we not prevent our children from hurting themselves and each other? If we should, then why would not a benevolent God interfere and prevent our own destructiveness?

Some theists hold that a certain amount of suffering is necessary to achieve moral and spiritual maturation. The obvious objection to this theodicy is that suffering often does not improve the human character.

Some would propose that "the devil" could be responsible for the evil in the world. However, if the devil has as much or more power than God, that would mean that we don't have monotheism; if he has less power, then he is not a coequal, and God remains responsible for what happens.

A possible explanation for the problem of evil, adopted by minimalist theists, is that God is not omniscient or omnipotent and is therefore unaware of the evil or unable to stop it. Process theologians like Hartshorne claim that God, just like the natural order, is in the process of becoming and cannot prevent evil in the world.

The argument that God has created the best possible world doesn't explain the need for so much suffering associated with natural disasters. The number and severity of natural disasters such as earthquakes, cyclones, flooding, volcanoes and tsunamis in recorded history is mind-boggling. In addition, man-made disasters and the high prevalence of disease have contributed to cause much additional pain and suffering. The Antioch earthquake (AD 526) killed between two hundred and fifty and three hundred thousand persons. In 1556, the Shanzi earthquake in China resulted in eight hundred and thirty thousand deaths. During 1775 the earthquake and tsunami that hit Lisbon caused about one hundred thousand deaths and prompted many people to question their religious beliefs. The Lisbon earthquake presented formidable difficulties for religious people. It occurred on All Saint's Day at a time when the churches were very crowded. Along with churches and cathedrals, the earthquake managed to wreck most of the city's monasteries and convents, and a disturbing number of sacred images were damaged in the process. Religious leaders set about the task of justifying God's action in Lisbon, and implicitly reconfirming his existence. Some miraculous survivals were credited to one saint or another. Protestants pointed out that the victims were predominantly Catholic, and the sinfulness of the city's population was blamed for the destructiveness of this event. Voltaire admitted to the difficulty of reconciling this horrible event to his view of a just and rational God.

In 1839 a cyclone caused three hundred thousand deaths in India, and in 1887 the Yellow River floods in China killed more than a million people. China also suffered two hundred and thirty four thousand deaths in the Haiyuan earthquake in 1920 and two million deaths in 1931 when the Yellow River overflowed its banks. In 1970 the Bhola cyclone killed half a million people in what was then East Pakistan. More recently, the 2004 tsunami of Indonesia caused about a quarter of a million fatalities and a similar number perished in Haiti during the 2010 earthquake. Thousands died during the 2011 earthquake and tsunami in Japan.

Violence, predation and death are the ways the natural world operates. Humans and animals have to kill other animals or plants in order to avoid starvation. In addition, humans are responsible for much of the pain and suffering in the world. We kill each other because of ideology, religion, greed, competition and status, and we are responsible for the mass casualties associated with war, terrorism and genocide. It is almost impossible to catalogue a comprehensive list of human atrocities perpetrated since the beginning of recorded history, including those taking place at the present time. Every year we allow thousands of children to starve to death or to become ill with preventable diseases. One third of the world population lives in extreme poverty.

One often wonders why there are so many diseases and why they have to cause so much physical pain and mental anguish. The prevalence of disease is universal and the suffering that it causes is familiar to everyone. The World Health Organization launched the Global Burden of Disease initiative in 1991. It assesses the most common causes of death, and the years of life lost to premature death, the years lived with disabilities and the disability adjusted life-years. The most frequent health problems mentioned in the last index were ischemic heart disease, lower respiratory tract infections, stroke, diarrhea, HIV-AIDS, malaria, low back pain, pre-term birth complications, chronic obstructive pulmonary disease and traffic accidents.

We suffer from a multitude of diseases including those due to infectious agents, neurologic disorders, genetic and environmental causes, immune mechanisms, hemodynamic and circulatory abnormalities, nutritional deficiencies and cancer. The wear and tear in our body causes the afflictions of old age. The cells in our bodies are susceptible to several forms of injury that includes ATP depletion, mitochondrial damage, calcium influx, free radicals, membrane injury causing increased permeability and damage to DNA and proteins.

Infectious diseases are highly prevalent and have been an important cause of death and disability in the world. Millions of people died during the Middle Ages due to the plague, an infectious disease caused by a bacterium. Smallpox and measles were introduced to the Americas by the Spanish explorers; these diseases killed thousands of people. Communicable diseases have been responsible for the death of millions of people since the beginning of the last century. The Spanish flu in 1918 killed between fifty and a hundred million people. Malaria and tuberculosis have killed millions of people around the world. These diseases are still not controlled and claim many victims every year in the third world. The AIDS epidemic started in 1981 and has already claimed more than twenty-five million lives worldwide, mostly in Africa.

Pain is one of the oldest scourges of humanity. It is associated with injuries, diseases and even physiological events such as menstruation and childbirth. The papyrus of Ebers, one of the oldest medical documents, deals with pain and with ways to manage it, and Homer relates that Helen used a pain-killing agent to care for survivors of the Trojan War. In ancient China and India, people used drugs and special powders to obtain relief from pain. Alcoholic brews and opium were the first agents to be used. The wild grape was domesticated in Egypt around four thousand BCE and alcoholic beverages became an ingredient of many pain-killing potions. Several varieties of poppy were grown in Egypt and Mesopotamia. The word opium means fresh juice from the poppy. In the Hellenic world opium was an upper-class indulgence; it was also used by Celsius, and to a lesser extent by Galen, to care for their patients.

Disorders of the nervous system have a great potential to cause pain and its associated suffering because the brain is the center of our sensory, emotional and cognitive abilities. The peripheral nervous system is required for the sensory and motor responses to environmental stimuli. Damage to the receptors or the cranial nerves may cause severe disability. Blindness and deafness are among the most devastating conditions that exist and they are prevalent worldwide. Paralysis of the facial nerve (Bell's palsy), which is usually idiopathic, may cause facial deformities. The carpal tunnel syndrome, associated with repeated trauma to the wrist, is one of the most common peripheral nerve disorders.

Many neurologic diseases are genetic, others are congenital or present at birth and others unfold as the individual grows. The best known genetic disorder is Down's syndrome, which results from an extra chromosome 21 and may be associated with severe mental retardation. The cause of autism spectrum disorders is unknown but experts believe that they have a genetic component. Other genetic diseases include the fragile X and Rett syndromes.

Other neurodevelopmental disorders are caused by infections such as rubella, syphilis, herpes simplex and toxoplasmosis. Lack of a vitamin, folic acid, may result in nerve tube defects like spina bifida or anencephaly, and iodide deficiency may cause cretinism. Metabolic disorders such as phenylketonuria may lead to mental retardation unless it is diagnosed early and treated with a special diet. The toxicity of alcohol and heavy metals to the fetus is well established. Traumatic birth injury is an important cause of cerebral palsy. Seizure disorders and multiple sclerosis are neurological diseases that usually affect young people in the prime of their lives. The latter is an immune-mediated disease that affects the myelin sheath that surrounds nerve fibers.

Traumatic brain injury may occur at any age as a result of accidents, fights or contact sports. Brain trauma results in fractures of the skull, concussions, contusions, hematomas, and alteration of consciousness ranging from mild confusion to deep coma. Coma has multiple causes and has to be differentiated from the locked-in syndrome and persistent vegetative state. In the former, patients don't move but they are able to answer questions moving their eyes. Patients in an irreversible vegetative state are unable to interact with the environment but appear to have cycles of sleep and wakefulness. Patients in deep coma may not breathe normally and may require artificial ventilation. The management of patients in coma often creates difficult ethical, medical and legal problems.

Repeated trauma as in boxing and football may produce not only damaged brain tissue but also may predispose to dementia. Injuries to the spinal cord in the neck may produce tetraplegia, while those below the thoracic spine, paraplegia. The patients with the latter condition suffer from, in addition to paralysis and sensory defects, uncontrolled sympathetic output, pain, hypotension, spasticity and bowel and bladder dysfunction. Protrusion of intervertebral disks and compression fractures of the vertebral bodies due to osteoporosis are common causes of back pain in older individuals.

Dementia afflicts a large number of elder persons and is becoming a common cause of death and disability now that people live longer. The clinical picture is characterized by a marked decline in cognitive ability in comparison to the baseline level of function. It is caused by Alzheimer's disease and other neurological conditions. Parkinson's disease also affects older individuals. It is associated with involuntary movements as well as rigidity and impaired voluntary movements.

Religion in the United States

In contrast to the rest of the developed countries, the United States appears to be a very religious country (Niose, 2012). Most surveys are difficult to interpret but approximately eighty percent of Americans currently identify themselves as religious; thirty percent of adults under thirty, though, are not affiliated with any religion. It is noteworthy that other wealthy and developed countries, especially in Europe and the Pacific Rim, have become less religious. Nonetheless, traditional religion in the United States didn't fade away, as many predicted, and many groups have flourished in the last three decades. Evangelical Christianity, an outgrowth of Christian fundamentalism, became popular in the 1940s and has continued to grow until the present supported by highly organized and influential conservative groups like the Moral Majority and the Christian Coalition.

Critics of religion in America point out that there is an obvious gap between high levels of apparent religiosity and the high prevalence of corruption, political dishonesty, tax cheating, sexual promiscuity, marital infidelity, violence and crime. Most fundamentalist religions have failed to heed scientific advances and, in contrast with mainstream churches, they cling to pre-modern narratives. Many individuals in these churches only follow some aspects of their religion and neglect others. For example, Catholics are generally pro-life but at the same time most of them practice birth control, which may also result in the loss of a potential human life. Many attend church regularly only because they like the rituals and the social interaction. Many Christians, especially evangelicals, belong to the religious right, whose members try to influence our political system and claim the moral high ground. They are very active proselytizing in third world countries. Oddly enough, some of these individuals have been found to be especially greedy, elitist and militaristic yet claim a special relationship with Jesus, who was one of the world's most prominent social reformers, an egalitarian and a pacifist.

The influx of immigrants to the United States was largely the result of religious upheavals in Europe that were fueled by the Protestant Reformation. The Puritans earned their name from the desire to purify society, inspired by Calvinist theology. Their desire for self-governing parishes, their skeptical view of human nature and their sense of destiny became part of our cultural legacy. At the same time, the Puritans accepted the need for the church-state separation and the concept of religious freedom.

Religious tolerance was strengthened in the late eighteenth century when the Constitution and the Bill of Rights were drafted and adopted officially by the United States. Of interest, many of our founding fathers favored religious freedom to reduce any established clergy's interference with politics. Some of them were Enlightenment deists who believe in a God who set the universe on course with natural laws and then left it alone. In the nineteenth century many religious groups thrived, especially the Baptist and Methodist branches of Christianity. Millennial sects proliferated, claiming that they could divine the End Times prophesied in the scriptures.

The popularity of the Left Behind series of novels and related media is an indication of the continuing appeal of millennial religion in America today.

The religious climate also propelled the Church of Jesus Christ of Latter-day Saints and its faithful, the Mormons. The Mormons trace the origin of their religion to Joseph Smith who claimed to have found sacred tablets describing how the lost tribe of Israel migrated to the New World. Smith and his followers were chased out of several states for their outside of mainstream practices, including polygamy. He was arrested in Illinois but before his trial a lynch mob stormed the jail and killed him. His successor Brigham Young then led the faithful to safety in the Salt Lake valley.

In addition to home-grown pluralism, immigration also fueled religious diversity. From the mid-eighteenth century on, successive waves of immigrants came from Catholic countries like Italy, Ireland and Poland. The Catholic–Protestant split has influenced our culture and has shaped partisan political loyalties until the present time.

The American experiment in religious freedom resulted in a remarkable growth of Evangelical religion. This is a branch of Protestantism that is deeply committed to the Bible as the only authoritative source of revelation, stresses the significance of the adult conversion (being "born again") and recognizes the value of evangelizing. Evangelicals in the North led an awakened Christian conscience against slavery in the decades preceding the Civil War. In contrast, Southern Evangelicals defended their position claiming that slavery was divinely ordained and justified in the Bible.

The Temperance Movement, the crusade against intoxicating beverages, became another example of evangelical politics. The creation of the Anti-Saloon League in 1895 became a formidable force in the fight against alcoholism. The league effort was credited for the passage of the eighteenth Amendment to the Constitution prohibiting the production and sale of alcoholic beverages. The amendment was repealed eventually because it fueled political corruption and organized crime. Evangelical were also active in support of populist movements, denouncing the various elites and demanding that the views of the majority of the population be heard.

Catholics comprise a third of Christians in the USA. One fourth of Catholics favor the literal interpretation of the Bible. Ignatius Press, the largest Catholic publishing house in America, publishes books by authors who are supportive of the conservative views of the Vatican. Their views on contraception may have contributed to overpopulation and the spread of HIV infection in the third world. Recently, growing competition from Evangelicals and a damaged reputation due to the child abuse scandal have considerably weakened the status of the Catholic Church.

Approximately seventy million Protestants are Evangelical or "born again" Christians. They are the largest growing religious grouping in the world and most of their new members are poor and ignorant people from the developing world. In the United States, Evangelicals have recently lost some of their political constituency because of their emphasis on promoting war and their failure to support social justice, animal rights and environmental protection.

Christian Zionists favor keeping Israel's settlements in Palestinian areas because according to their notions this will facilitate the second coming of Christ. Perhaps even more dangerous to world peace is the Chalcedon Institute, which is dedicated to instituting Biblical law as the main governing force in the world. These Reconstructionists hold that believers should work towards achieving God's kingdom on earth in the here and now, not after the second coming of Christ.

About three in four Americans associate with some form of Christianity. A large group of Christian churches compete with each other; this pluralism assures an abundant supply of religious options. There are four dominant Christian traditions in the United States: evangelical, mainline protestant, African American and Roman Catholicism. Today, the largest group is the evangelical tradition. There are hundreds of evangelical denominations, the largest of which is the Southern Baptist Convention. This tradition became popular after the presidential candidate Jimmy Carter announced that he was a "born-again" Christian. The impetus for its growth came from student organizations, from national groups and from the endorsement by several politicians.

In addition to white members, black Baptists and Pentecostals are fervent evangelicals. Moreover, improved racial relations have enabled more interaction between them. Nevertheless, a considerable gulf remains between black and white evangelicals. The former tend to be more politically liberal than the latter. Evangelicals also differ in their interpretation of millennial events, their belief in the inerrancy of the Bible and their emphasis on the availability of the gifts of the Holy Spirit such as speaking in tongues, faith healing and prophecy. A recent survey found that only nine percent of Americans under age thirty are evangelical. Political participation has increased considerably, however, and today most evangelicals are conservative and participate in political activities.

Mainline Protestants include many groups like Methodists, Episcopalian, Presbyterian, Lutheran, northern Baptists and members of the United Church of Christ. They are also called "liberal Protestants" with the adjective liberal referring to the religious not the political orientation. They believe that the Bible is not literally true. They embrace Darwinism and other scientific advances. Some of them are politically liberal, a movement that probably started when they joined the progressive social causes in the 1960s. Mainline Protestantism has declined in the last few years perhaps because their religious beliefs and the conflict between lay members and the national leaders.

Catholics used to shield themselves from the dominance of the Protestant culture. This situation no longer obtains and Catholics are fully integrated socially and politically. As a universal institution, the Catholic Church focuses not only in the American church but also in the global and hierarchical structure of the church. As a result of Vatican II (The Second Vatican Council, 1962–65), the Catholic Church became more democratic and for the first time Protestants were accepted as fellow Christians. The Mass was thereafter celebrated in the local language and the bishops of every country were given more authority to speak on behalf of the church in their respective lands. Since Vatican II, church attendance has decreased; some individuals have moved to Protestant churches and there is a chronic shortage of priests. An important challenge for the Roman Catholic Church has been the discovery that many dioceses had failed to discipline priests that were engaged in the sexual abuse of minors.

There is a very small number of Americans who have a non-theistic conception of God that gets around the problem of evil and doesn't conflict with science. Deists view God as a cosmic force that created the universe and its laws but doesn't intervene in the world. In this view, God could have created the world, setting up the laws of nature and leaving them alone to do their work. These laws may have the capacity to create life forms and facilitate evolution along creative pathways. Pantheists like Spinoza see God as the totality of all there is. They deny the existence of a personal god and the notion that God is ontologically different than the world. Panentheists hold that the universe is part of God but God is more than the universe. In addition, there are believers who embrace process theology or assert that God is ineffable, the images of God in human terms are inadequate and represent naïve attempts to know the unknowable.

Many twentieth -century Christians, like Tillich and Bultmann, embraced liberal theology. They have distanced themselves from the concept of a real, transcendent God and affirmed that religious statements are mainly symbolic. The Unitarian-Universalist church is a liberal church that draws inspiration from many religious and philosophical traditions including Atheism, Buddhism, Christianity and Paganism. Liberal Christians compete with evangelicals, mainline Protestants, African-American churches and Roman Catholicism for the leadership of the Christian church and its relevance in the modern world. Some Jewish theologians like Lerner and Kaplan have also embraced liberal theology.

Many Americans don't have a particular religious affiliation. A small minority identify themselves as atheist or agnostic (Martin, 2007). Negative atheists have no specific beliefs; positive atheists deny the existence of any deity. Some atheists only deny the God of theism but believe in other deities. Believing that there are no deities is difficult to prove because it is not possible to prove a negative statement, especially the existence of entities that may be outside the natural world. Agnostics are usually seekers who prefer to suspend judgment about God's existence. Some prefer to call themselves skeptics, rationalists, infidels, free-thinkers or secular humanists; the last ones prefer to address the existential and ethical issues facing humanity rather than to focus on metaphysical questions. They are concerned that religious organizations have too much money, too much power and too much influence in the political process.

It is not well recognized that atheism, like theism, has an ontological and an epistemological dimension. Atheists may address the ontological question of whether or not God exists or the epistemological question to the extent of which God's existence is or can be known or justified. The ontological atheist or theist is in a weaker position than those who hold both an ontological and an epistemological position. Atheists claim that the burden of proof should fall on those who claim that God exist, not the other way around. In addition, they argue that although absence of evidence is not necessarily evidence of absence, a striking lack of evidence for the existence of an entity suggests that it doesn't exist. Some admit that they cannot prove that is impossible that God exists; they just claim that it is improbable.

Atheism has a long history. The pre-Socratic philosopher Xenophanes expressed disdain for those who worship anthropomorphic deities and Socrates, many years later, was sentenced to death for rejecting some of these deities. Protagoras, a contemporary of Socrates, was an agnostic who was persecuted for his ideas. Carneades of Cyrene denied that the universe was part of a divine plan. Democritus presented an account of the world's origin devoid of any religious explanations. The idea of moving atoms in a void was also proposed by Epicurus. His followers embraced a form of religious materialism with its anti-religious bias and its concern for avoiding pain and seeking happiness. In Rome, his leading advocate was the poet Lucretius who described the Epicurean view of nature with great precision. The advent of Christianity as the dominant belief system in the West eclipsed other alternative views including atheism.

Western non-belief reappeared in the seventeenth century with the rise of modern science. The writings of Bacon and Bayle opened the way for materialism and the rejection of religious dogma that characterized the eighteenth century Enlightenment. Hobbes articulated a view of materialism that left very little room for God. Meslier was a priest who wrote a long defense of atheism but kept it secret during his lifetime. When his book was finally published, it became a source of inspiration for anti-clerical writers like Voltaire and atheists like d'Holbach. Hume didn't declare himself an atheist but claimed that miracles violate reason and attacked the traditional arguments for the existence of God. The rise of the empirical science combined with an anti-clerical view influenced thinkers like Feuerbach, Marx, Nietzsche and Freud. Darwin's theory of evolution by natural selection called into question the belief that humans have a divine origin.

In the twentieth century, many philosophers criticized religion. Russell defended atheism and agnosticism and Ayer argued that the discourse about God is not verifiable or self-evident, and therefore meaningless. In the Continent existentialist philosophers based their denial of God on their appraisal that the universe is absurdly pointless and their conviction that human freedom would be compromised by the existence of a deity. In the United States, Twain and Ingersoll attacked the claims made by religion and contributed to the spread of atheism. The most recent wave of atheists has been dubbed the "New Atheists." They include writers like Dawkins, Dennett, Hitchens, Harris and Stenger. This movement is noted for its militant and polemic condemnation of all religious belief.

Some contemporary atheists prefer a worldview based on physicalism, reductionism and ontological naturalism. They hold that the universe had no beginning; it has always existed. They believe that life may have originated when self-assembling and self-replicating macromolecules acquired a limiting membrane and a source of energy. Living organisms evolved by natural selection creating all the numerous species that now exist. Eventually, multicellular organisms acquired neurons and the progressive organization of neurons in the brain created a complex and distributed network capable of cognition and mental experiences.

The largest group of secular Americans includes those who have no religious identity, don't identify themselves with any religious group, and have no place for God or religion in their daily life. They describe themselves as spiritual to one degree or another. This is signaling a movement from organized or communal religion to individual and self-directed approaches to religion. Recent surveys identify these individuals as "Nones," people who claim no relation with organized religion. (Another factor that may be associated with the increase in the number of "Nones" is the growth of "New Age" religions.) Many observers believe that secularism, especially among the millennials, will increase in the future and they may become essential actors in the political system.

Religious naturalists find special meaning and value in nature or some aspect of the natural order. They prefer to develop a spiritual dimension of naturalism not explicitly tied to any particular religious tradition. Many of them believe that the natural sciences can provide objective and reliable descriptions of nature. Others, like Dewey, go beyond the scientific disciplines in developing their views about nature and its processes.

To deal with existential concerns, people without religious affiliation acknowledge that they are part of the natural world, with its moral ambiguities. They consider that humans have a dual nature; they have emotions but at the same time they realize they are endowed with the power of reasoning. They accept that humans are not fully autonomous beings and their actions may be controlled by unconscious processes and hidden cultural and biological influences, so that free will may be an illusion. They don't believe that souls exist or that there is an afterlife; they believe that this is the only life we have, and we should follow humanistic principles to give meaning to our lives and alleviate suffering in the world.

The media and politicians tend not to address people without religious affiliation. Politicians don't associate themselves with secularism or admit that they are nonbelievers, for fear of losing their constituency among the majority of Americans who are religious believers. Nevertheless, secular Americans are generally just as patriotic and moral as the rest. While they support the separation between Church and State, most don't seek to undermine religious beliefs that help others get through life and are the basis of their cultural identity.

Secular Americans would like to be counted and respected; twenty percent of Americans are unaffiliated with a particular religion. They would like to have religious freedom but also freedom from religion, that is, freedom from coercion or proselytization. They believe that the continuance of religion should not depend on the indoctrination of children, who are credulous and believe and obey their parents, teachers and the expectations of society. If there is to be any religious instruction, in their view, it should be started in high school or college, at a time when individuals are mature enough to rationally evaluate what they are being taught.

CHAPTER 12. SCIENCE AND ITS LIMITATIONS

During the Middle Ages many of the scientific advances made by the Greeks and Romans were not available to Europeans but were preserved by scholars in the Islamic world. Only the revolt against ecclesiastic authority brought on by the Renaissance and the Reformation made it possible for science to flourish in the West. Protestants encouraged scientific discovery based on their work ethic and their belief that God is revealed by the study of nature. During the Enlightenment, a clear distinction was made between science and religion. An intellectual awakening did not take place in China or the Islamic countries at that time because scientific work was constrained by government and/or religion.

Prior to the scientific revolution science and philosophy were based on the opinion of authority figures. Scholasticism, a mixture of Christian doctrine and Aristotelian philosophy, was challenged by the discoveries of Copernicus, Galileo and Newton. Rationalists like Descartes held the primacy of reason to obtain knowledge about the world; their views were opposed by those of the Empiricists who claimed that experience, not reason, was the best way to understand reality. The latter view, however led to Positivism and the elimination of scientific doctrines that could not be verified by observations.

The Logical Positivists claimed that there are only two kinds of meaningful statements: the analytical, such as mathematical or logical and the empirically verifiable like the statements of science. Their theories ran into trouble and they had to admit that many scientific theories are not conclusively verifiable. There were three responses to the failure of Logical Positivism: Instrumentalism, Inductivism, which weakened the original criteria of verification and Deductivism, which was advanced by Popper and replaced verifiability as a criterion of meaning with falsifiability as a criterion for the scientific. Popper rejected the view that scientific knowledge is derived from facts by inductive inference, and concluded that hypotheses that are tentatively proposed to accurately describe some aspect of the world must be falsifiable before they are granted the status of a scientific theory. Scientific truths are tentative; they can never be proven to be true, they, however, can be proven to be wrong or falsified.

Kuhn held that science progresses not by the accumulation of well-confirmed truths or the elimination of conjectures shown to be false, but by revolutionary upheavals. In his book entitled "The Structure of Scientific Revolutions" he proposed that science doesn't progress by the steady addition of knowledge but by revolutionary steps due to the introduction of new paradigms. As a field of study matures, anomalies arise that accumulate and create a crisis that leads to the formulation of a new paradigm. This book advanced the possibility that the new paradigms of science may not be progressing enough for us to formulate a true representation of the world.

Feyerabend held that there is no scientific method, that the appeal to "rationality" depends on propaganda and that science is not superior to other disciplines such as astrology. Lakatos was concerned with the question of demarcation between science and non-science. He claimed that the hard core of scientific theory cannot be broken by observation alone, only by an alternative theory. Quine statements regarding underdetermination gave encouragement to cynics and Van Fraasen "constructive empiricism" held that the goal of science is not to arrive to explanatory theories but only to arrive to empirically adequate theories. Recently, Haack has taken a middle ground position. She asserts that science doesn't have a privileged epistemological position. According to her scientific progress may be gradual or revolutionary, and may involve belief revision or conceptual change, accumulating truths or repudiating falsehoods (Haack, 2003).

Most people claim that the scientific method begins with observation, moves to theories and then produces generalizations with predictive ability. The problem with this view is that our knowledge and our expectations influence our observations and that it relies on induction rather than deduction. In deductive arguments if the premises are true the conclusions must be true, the conclusions of inductive arguments with true premises may or may not be true. Another important non-deductive argument used by scientists is known as Inference to the Best Explanation or abduction. In this type of argument we judge the plausibility of the hypothesis in terms of the sort of explanation that it offers.

One response to the problem of induction is that its application has been reasonably fruitful, that it has led to many advances and that it works most of the time in predicting the future behavior of the natural world. Popper suggested a method to get around the problem of induction. He argued that scientists don't begin with observations, they begin with a theory. These conjectures are then subjected to experimental testing. Testing is aimed not so much in proving that the conjecture is true, but rather to falsify it, to prove that it is false. Thus science progresses by means of conjectures followed by refutation. His theory may be historically inaccurate in that it doesn't account for many significant developments in the history of science.

Science enjoys a peculiar epistemological authority because it uses a uniquely objective and rational method of inquiry. Science progresses by accumulating truths or near truths confirmed by empirical evidence or by testing theories against basic statements. False conjectures can be replaced by corroborated theories. The core standards of good evidence are not the privilege of science; they apply to other disciplines of knowledge and scholarship.

That being said, science can also be mis-used. Scientists are members of society; they are not isolated. Thus, they can be subject to political and cultural influences, and their work can be directed to fit someone's goals. Further, any invention, such as television or explosives, can be used for positive or negative purposes. That is a separate problem. The fact is that science can show a basis for the conclusions it has reached; it does not rely on faith.

Many people have noticed the complex landscape in which scientists develop their theories. The history of scientific discovery shows that experience doesn't necessarily lead to new scientific theories; cultural influences, conjectures, peer criticism and alternative explanations often play a major role in the formulation of scientific theories. Postmodernists even claim that most scientific ideas are arbitrary, conjectural and lack justification; they are only stories or "narratives," as we cannot prove such a thing as an "absolute truth." In spite of these criticisms, scientific theories have been shown to have an important predictive value and have created substantial progress. Our incomplete representations of reality have helped us to better understand the world.

Many scientists are ontological naturalists. Some of them claim that all entities and processes are physical and can be reduced to simpler components, everything is part of the natural world, and supernatural entities don't exist. Others, disturbed by the fact that we can provide only partial evidence and support for this position, restrict themselves to methodological Naturalism, which stipulates that hypotheses should be tested by scientific methods and explained solely by reference to natural causes and events. Technically, we cannot prove that supernatural entities don't exist. We cannot prove a negative. Naturalism's ethical dimension is associated with humanism;

it deals with existential questions, morality and other problems facing humanity.

Most scientists generally consider that humans have no way of knowing the ultimate reality, and indeed, the idea that humans can or should understand everything in the universe is an ambition, not a reality.

The task of scientists is to look at the world using the best possible methods, and to continually seek new ways of discovering how the world works. Most often they generate hypotheses or explanations that predict how the natural world works and then test the hypotheses, trying to confirm or reject these ideas using experimental methods and evidence obtained in the real world.

We still don't understand many important physical features of the natural world. We don't know the link between the abstract micro-world of quantum physics and the subjective macro-world of daily experience. Some of the basic questions for which science doesn't have a definitive answer are shown in Table 2. Science has many limitations, beginning with the problem of induction and the objections to scientific realism. The physical sciences don't give a full understanding of the origin of the universe; the biological sciences have not explained how self-reproducing life forms originated from inanimate matter; and the psychological sciences are only exploring what we call "human nature" via electrical patterns in the brain that result in consciousness and phenomenal experiences. And some questions like free will, morality and the meaning of life simply may not be susceptible to universal claims of right and wrong, true or false. These remain in the area of philosophy, for lovers of wisdom to ponder and discuss in an open-ended examination.

TABLE 2. BASIC QUESTIONS FOR WHICH SCIENCE DOESN'T HAVE A DEFINITIVE ANSWER.

The problem of induction
The creation of the universe
The origin of life from inanimate matter
Consciousness and phenomenal experiences
Complexity and the nature of emergence
The weirdness of quantum events
The fine tuning argument
Dark matter and dark energy
Free will and morality
The meaning and purpose of life

Scientific advances have been aided by the invention of instruments capable of overcoming the limitations of our senses. For example, the spectrum of human vision extends down only as far as one or two tenths of a millimeter, the realm of the very small goes largely unnoticed by us. The introduction of the optical microscope permitted scientists to visualize cells and bacteria. To visualize viruses and molecules smaller than a single light wave scientists had to wait until the electron microscope was invented in the twentieth century.

In spite of advances in instrumentation, scientists still don't completely understand the nature of the basic building blocks of the universe and the laws that govern them. Some ancient philosophers claimed that matter could be divided into smaller pieces indefinitely. Others, like Democritus, argued that matter consisted of small, indivisible entities called atoms. Modern physics began with the formulation of the atomic hypothesis at the end of the nineteenth century. According to the theory, atoms are made of three types of smaller particles: protons and neutrons in the nucleus, and electrons that orbit around the nucleus. The nucleus ties most of the mass in an atom, mass is a concentrated form of energy and the two are linked by Einstein's famous equation.

The twentieth century witnessed several paradigm shifts in physics and mathematics. Einstein introduced a new concept of space and time that conflicted with Newton's theory (Chown, 2006, Stenger, 2003). By combining the laws of gravity and motion Newton was able to explain the movement of celestial bodies. His theory was based on absolute space and time. According to Einstein's special relativity, though, nothing can travel faster than light; our measuring sticks have to adjust to conform to this physical fact. Time intervals do not in themselves have absolute meaning but rather depend on the state of motion of the observer who measures them. Einstein's equations of general relativity expressed the link between the four-dimensional space/time curvature and the gravitational properties of material bodies. It is still unclear, however, whether space and time emerges from something more fundamental like a holographic projection or a network of intersecting quantum threads.

Quantum theory explained how energy comes in discrete packets and the wave-particle duality found in the double-slit experiments. Quantum physics has forced physicists to abandon the view that nature is deterministic. It describes a probabilistic universe, a reality which depends on "being observed" and "spooky actions at a distance." The latter appears to undermine Einstein's theory of relativity.

In spite of its successful application, we still don't have a definitive interpretation of quantum physics. The Copenhagen interpretation suggests that the wave function of possibilities "collapses" to an actual event. The "many-world" interpretation proposes that that the universe splits at each quantum measurement into two parallel universes and the Bohm interpretation proposes that quantum behavior is fully deterministic but unknowable.

The Standard Model of particle physics has been a very successful theory in providing a description of the particles and forces that make up the universe but cannot be considered a grand unified theory and doesn't include gravity. According to the Standard Model particles come in two types, those that make up matter, known as fermions, and those that carry forces, which are known as bosons. Protons and neutrons in the nucleus are made of quarks; the electrons are also elementary particles; ordinary matter is made of electrons and quarks.

The Standard Model claims that all forces in nature arise from the exchange of a particle. The familiar electromagnetic force is carried by photons, the carrier of the strong force is the gluon and the weak force is carried by two particles called W and Z. The detection of gravitational waves would suggest that gravity is a quantum phenomenon and gravitons exist. The discovery of the Higgs-like boson has provided additional support to the model.

According to quantum field theory the different particles arise from fields that permeate the universe, every particle represents a vibrating wave in a particular field. Thus the photons that carry the electromagnetic force are vibrations in the electromagnetic field. The Higgs field that creates the Higgs boson behaves different than other fields in that it has less energy when it is non-zero than when it is zero. It provides a medium through which other particles move and influences their behavior. The physical order is difficult to define; quantum fields may be the deeper reality while particles are simple excitation of these fields, the order in which particles and space-time emerge from energy fields, or fields and particles may constitute a unity, they don't exist as separate realities.

Many physicists have attempted to create a grand unified theory that includes the four known forces of nature, gravity, electromagnetism, the weak nuclear force and the strong nuclear force. Quantum electrodynamics was proposed by Feynman in the 1940s. In the 1970s the weak and the electromagnetic forces were united. Supersymmetry (SUSY) is a theory that holds that force particles and matter particles are two facets of the same thing; it unifies the non-gravitational forces of nature. M-theory may unify all forces in nature; it results from joining supergravity, the best known supersymmetric theory, with string theory. M-theory posits that everything is made of tiny vibrating strings and membrane-like objects in an eleven dimensional universe. Only four of the eleven space/time dimensions are observable; the rest must be rolled up to a very small size.

Recent discoveries have shown that the Standard Model accounts for only about five percent of the fabric of the universe. "Dark matter," detected by is gravitational effects, holds the galaxies together; it makes up about twenty seven percent of the mass/energy of the universe. It may consist of "super-partner particles" described by Supersymmetric theory; however, these particles have not been identified in recent experiments using the Large Hadron Collider. "Dark energy" makes up sixty eight percent of the mass/energy of the universe. For several billion years this repulsive gravity has dominated the cosmic density and caused expansion of the universe to grow exponentially. It is unclear if it is a property of space (Einstein's cosmological constant) or something in space. The strength of this force will determine if the universe will contract or expand forever carrying all the galaxies we now see out of our view.

The Standard Model of particle physics and cosmology is incomplete and scientists are unable to explain the ultimate origin of the universe. Scientists also don't have an explanation for how life originated spontaneously from non-life and have trouble explaining how phenomenal experiences emerge from neural networks. They have to look for ways to find value and meaning in this life, discover ways to assuage existential angst and behave morally in the absence of free will and religion. In addition to the aforementioned problems, scientists have to deal with the epistemological problem of how to acquire knowledge, the metaphysical issue of scientific realism and the limitations of scientific determinism and reductionism.

Truth and Knowledge

Epistemologists define knowledge as justified true belief. Nonetheless, the Gettier problem suggests that there are examples that show that we can have knowledge without one of the three components of its definition. To have knowledge we need a deeper understanding of the factors that influence the degree of justification in believing in something. According to evidentialists what makes a belief justified is the possession of evidence, for reliabilists justification is based on the use of reliable cognitive processes.

The Platonic view is that the truth is clear, certain and knowable; to distinguish knowledge from belief we need absolute, eternal first principles derived through reason (a priori). His view was foundationalist; it held that the structure of knowledge is like that of a building, some beliefs that support other beliefs don't require justification. In the Aristotelian perspective the distinctions between truth and knowledge are a matter of degree, not absolute. Knowledge requires empirical concrete evidence (a posteriori), rather than a strong foundation, it focuses instead in how well beliefs fit together.

Rationalists argued that knowledge is based on reason; they leaned heavily on a priori arguments. On the other hand, empiricists claimed that knowledge derives from experience, they used a posteriori arguments. Pragmatism, defines truth and knowledge in terms of what works, without requiring a secure basis for this knowledge. Instrumentalists assert that beliefs are true if they allow us to do things which we could not do unless we took them to be true. Postmodernists deny that we can have absolute knowledge or possess self-evident truths.

Skepticism points out that we may be wrong about even our more confident beliefs about the world. Skeptics rightly point out that an absolute standard for truth and knowledge cannot be attained. We have to accept that there is always a gap between absolute certainty and the high degree of probability that we seek. It is an old tradition that goes back to the Greek philosophers. The Ancient Greek Pyrrho was the most extreme skeptic of all time. Descartes started by doubting everything until he came to the conclusion "I think, therefore I am," and Kant held that we cannot know the world beyond appearances. Hume was skeptical about causal reasoning, but he clearly believed that we can identify causation and understand how the world works.

Realism

Metaphysical views relate to the theory of knowledge and how we acquire knowledge of our surroundings. Common-sense realism is the belief that things are in reality as they appear to be in the mind. It is the view that there is a mind-independent reality and that things are more or less how they appear to us. This theory doesn't stand up well to arguments about the reliability of sense perception. Representative realism gets around this objection by suggesting that all perception is the result of awareness of inner representations of the outside world. Scientists realize that our senses are limited and our perceptions of this reality are created in the brain; there is no way to remove the observer from our perception of the world. We perceive the external world through our senses and what we perceive is only the fraction of the real world for which we have a sense organ. Representative realism provides a response to critics of naïve realism. For example, the color red doesn't exist, we experience color as our brains interpret the particular portion of the electromagnetic spectrum scattered by objects. A color-blind person might well see the object in a different way. On the other hand, representative realism makes the real world unknowable.

Idealism gets around the problem by asserting that there is no justification for saying that the external world exists at all because it is unknowable. Kant transcendental idealism holds that there is a mind-independent reality but we cannot experience it as it is in itself, our world is confined to the phenomenal world of appearances. Even the latter may represent the way our minds work rather than the way the world is. In addition, language not always represents the world with complete accuracy; Wittgenstein asserted that the meaning of words is always to be found in their use. A major criticism of the idealist theory of perception is that it may lead to solipsism, the view that all that exists is in the mind.

Phenomenalism agrees that we only have access to sense experience but claims that experiences can be described in terms of patterns of actual and possible sense experiences. For them the possibility of a visual experience continues even if we are not looking at the object. Causal realism assumes that the causes of our sense experiences are physical objects in the external world. It also assumes that the beliefs that we acquire from our sense organs are generally true and were selected by evolution in order to give us a reliable knowledge about the environment.

The scientific method to obtain knowledge was championed by Bacon who proposed that the process has to be inductive and not deductive. As mentioned before, Popper stressed the problems with induction and held that scientific theories are conjectures that have to be subjected to experimental testing. We cannot know for certain that any theory is absolutely true, any theory can be falsified. Kuhn, however, maintained that there is no single scientific method and that science creates paradigms that help structure inquire. When anomalies accumulate a scientific revolution takes place and a more fruitful paradigm ensues.

Scientific realism adopts a model-dependent view of reality. With the information they get from their senses or their instruments, scientists build testable mathematical models of reality or follow a general mode of inquire that is oriented toward solving specific problems (Hawking and Mlodinow, 2010). The no-miracles argument asserts that the truths of scientific theories are the best explanation for their empirical success, and if this was not the case, their success would be a miracle. Structural realism, held by some scientists and philosophers, claims a central role of relational aspects over object-based aspects of ontology.

Some scientists claim that quantum mechanics has done away with the concept of matter by holding that fields represent the deeper reality, whereas others hold that fields and particles don't exist as separate realities.

The features of the subatomic world described by quantum physics, like the wave-particle duality, quantum tunneling and entanglement are counterintuitive, probabilistic and alien. Quantum theory questions our concept of reality; physical variables may not have a specific value unless and until we measure them. This has propelled some people to speculate that Quantum theory supports the possibility that the human mind has the ability to control reality. Entanglement, the observation that measuring a property of a particle instantaneously changes the state of another particle far away, challenges our concept of locality.

Some theist and non-theist spiritualists claim that quantum theory argues against reductionism and leads us to embrace holism. This belief is based on the wave-particle duality and the collapse of the wave function, a phenomenon that suggests that the observer plays a constitutive role in the making of the physical world. Some individuals envision a universal cosmic consciousness that includes the human mind, and the belief that our thoughts can change reality and the course of our lives. Quantum entanglement and the possibility of superluminal transfer of matter have also been used by spiritualists to support the holistic argument. In spite of these claims, the Standard Model of particle physics, which is based on reductionism, continues to be fully supported by experimental studies.

Determinism

Newtonian mechanics is deterministic; it holds that if we know the position, velocity and forces acting on every component particle, the future of a system can be predicted. Chaos theory is deterministic because it posits that if the precise initial conditions were known the future state of a system could be predicted. Quantum mechanics, though, is indeterministic, it asserts that the results of measurements can only be described probabilistically (Mlodinow, 2008). It describes true stochastic or random events. According to quantum physics, nature doesn't dictate the outcome of any process; it allows a number of outcomes, each with a certain probability of being realized. The uncertainty principle states that if we multiply the uncertainty in the position of a sub-atomic particle by the uncertainty in its momentum (mass x velocity) the result can never be smaller than a certain fixed small quantity called Planck's constant.

The existence of physical structures in the universe reflects the tension between local order that results from the laws of physics and overall disorder associated with chance and contingency. The Second Law of thermodynamics makes the formation of complex structures improbable, but not impossible. It appears that the interplay of entropy and favorable physical laws has permitted nature to move in the direction of creating complex structures.

The evolution of hierarchically layered structures has been stochastic and entropic; meaning that creation of order has been selective and occurred in some places at the cost of increase entropy elsewhere.

Reductionism

Reductionism began to flourish more than three centuries ago when Newton formulated the law of universal gravitation which allowed him to predict the motion of celestial bodies. Since then, the view that the workings of the universe are reducible to a few fundamental physical elements animated by a regular and predictable behavior, has dominated naturalistic thinking. Reductionism has been highly successful in unlocking many of the secrets of nature such as the structure of the subatomic world, the mechanism of heredity and the evolution of the universe.

The formulation of the "uncertainty principle" of quantum mechanics and the understanding of chaos introduced important challenges to Newtonian mechanics. In chaotic systems even miniscule uncertainty about the initial values can result in a huge change in the long-term outcome of the system. That is why long-term weather prediction is accurate only to about one week into the future. Studies of logistic maps have resulted in the discovery of universal characteristics of chaotic systems.

Some scientists feel that in complex systems certain events cannot be derived from or reduced to basic elemental components (Kaneco, 2006; Clayton, 2006). Complex systems is a field of research that attempts to explain how a large number of relatively simple individual components, following simple rules and without the benefit of a central controller, evolves into a collective whole that creates patterns, uses information and in some cases adapts to the environment. Typical examples of complex systems include the economy, the brain, insect colonies and the World Wide Web. Because these systems lack a central controller they are called self-organizing, and because in these systems simple rules produce complex behaviors they are referred to as emergent.

Cellular metabolism is a complex system that is described as self-organizing. Order in this system is created out of disorder defying the second law of thermodynamics. According to Boltzmann, an isolated system will be in a more probable macrostate than in a less probable one. Unless work is done, disorder will always increase until it gets to a macrostate with the highest amount of entropy. Life forms, though, are open systems and create local order at the expense of increasing global disorder.

Shanon's mathematical solution to the problem of how to transmit signals rapidly and reliably over telegraph and telephone lines established the foundations of information theory. His work established the concept of information to characterize order and disorder. He defined information content in terms of the entropy of the message; a message could be any unit of communication, a word, a sentence or a single bit. The entropy of a source is defined in terms of message probability not with the meaning of the message. Information is processed via computation. In a biological system computation is what a complex system does with information to adapt to the environment. How information takes meaning in the absence of a central controller is a profound mystery.

The concept of emergence has been introduced to explain how new forms of organization emerge that are not predictable or explainable in terms of lower level laws, particles and forces. It describes top-down causation in complex systems, meaning that the whole may have a reciprocal causal influence on its own parts. Some theorists claim that no new laws are required to explain higher-level phenomena; the new principles are consistent with, or in principle derivable, from laws governing the lower level. Emergence could explain certain physical processes, the biosphere, the economy and perhaps consciousness.

A network is a collection of nodes connected by links. Network thinking focuses on the relationship between entities rather than the entities themselves. The existence of largely separate type-knit communities in networks is termed clustering. The number of links coming into or out of a node is called the degree of that node. High degree nodes are called hubs. A major discovery of network science is that high clustering, skew degree distribution and node structure seem to be characteristic of natural and social networks. Two models have been evaluated, the small world and the scale-free networks like the World Wide Web. All scale-free networks have the small-world properties, but not all networks that have small-world properties are scale free. In technical terms, this means that when the distribution of values is plotted over a wide range of values, the shape of the curve obtained remains identical.

It was found that in small-worlds the average path length is automatically reduced by rewiring the system by changing a few of the nearest-neighbor links to long distance links. An example of small world properties includes the nervous system, with nodes being neurons, and links the connections between neurons. Other examples include gene regulatory mechanisms, ecological systems, electrical grids and airline schedules. It is unclear why evolution would favor networks with small world properties. Researchers have suggested that a scale-free degree distribution allows and optimal compromise between local and global processing. This arrangement may save energy by avoiding the need to send signals over large groups of connections.

Small-world connections also favor synchronization, for instance, when a group of neurons fire simultaneously.

In spite of the aforementioned advances, attempts to derive a rigorous mathematical framework that explains the common properties of complex systems have failed. Prigogine's theory of dissipative structure, Wiener's cybernetics, Bertalafny's general system theory and Maturana and Varela's notion of autopoiesis all have failed to provide unified and universal laws of complexity or new ways to conceptualize complex systems.

The conflicting views of science and religion are evident from the information presented in the last two sections. Both views fall short of having all the answers. However, science makes continual progress. There may always be aspects of reality that we are unable to explore, but partial evidence and partial knowledge is still a better guide than faith, in the absence of any evidence.

CHAPTER 13. FALSE DILEMMAS

Making a choice between two options is difficult and may be subject to many pitfalls. We often make intuitive decisions and fail to consider the pros and cons. We frequently use the confirmation bias — our tendency to gather evidence that confirms previous expectations — to influence our decisions, typically by pursuing supporting evidence and dismissing contradictory evidence.

It is also possible that the two choices are partial truths. In the ongoing debate between science and religion (naturalism versus supernaturalism), we have been led to believe that there are two clear options: believing in the existence of God or claiming that nature is all there is. These two traditional options, as we have seen in the last sections, have many shortcomings. They represent a false dilemma or false dichotomy because there are other options available. For example, Deists believe in a deity that created the universe but no longer interacts with it, and Pantheists see God as the totality of nature. Methodological Naturalists embrace the scientific study of nature without assuming that everything is physical or that supernatural entities cannot exist and agnostics hold that we cannot prove or disprove the existence or the non-existence of God.

In the United States, a majority of the population has grown up under one of the three related Judeo-Christian-Muslim traditions and they have accepted the religious view to understand the world. Thus most individuals embrace supernaturalism, often without giving it much thought; they believe in souls and their immortality, free will and God-given morality. Many reject evolution and affirm creationism or intelligent design. Of course, many people put knowledge and reason ahead of faith and tradition. These individuals embrace science and naturalism to understand reality; they are usually atheists, agnostics or secular humanists who question God-given morality, immortality, the existence of souls or free will, and they seek further understanding through scientific realism and logical theories like evolution by natural selection.

Traditional religion endures because it seems to provide an answer to the mysteries of life, assuages existential fears, helps us to face mortality, promotes social cohesion and gives a ready code of morality. Religious groups have also brought spiritual comfort, education and health care to many people. The Judeo-Christian tradition has been the centerpiece of culture in the West. It has inspired great works of art, like the Chartres Cathedral and Bach's music. We are accustomed to using its rituals to mark the most important events in our life including birth, marriage and death. Because religion has been inextricably interwoven into our culture, it will take a long time for believers to evolve, to become scientifically literate and to modify their beliefs to make them compatible with the current state of knowledge.

One important reason for the persistence of religious beliefs is that we have religious freedom but we don't have freedom from religion. We are pressured to adopt the religion prevalent in our country of birth and our family of origin. This occurs at an age in which we are credulous and unlikely to question the beliefs of the prevailing authority. It is usually very difficult for people to leave their original religion. Apostasy is treated by rejection or disciplinary actions by society and/or the threat of punishment by God in this life or in the afterlife. Throughout history, religion has created powerful institutions that have exerted rigid control over its members and has severely castigated those who embrace other views, calling them heretics. Yet there are thousands of religions, with different Gods, and all of them cannot be right.

As we previously mentioned, the proofs of God's existence offered by religion are flawed and contradict the findings of science. Many people attach too much importance to this question, a question that we cannot resolve. There is no requirement for human beings to have all the answers. There are limits to human knowledge, as uncomfortable as that may be.

The advances of genetic engineering, evolutionary theory and molecular biology undermine many of the tenets of traditional religions. The miracles described in religious texts seem capricious and violate the laws of physics. Some of the events described in the Judeo-Christian tradition (the flood, the virgin birth and the ten commandments) have been shown to have been picked up from earlier cultures.

Some fundamentalist religions favor self-deception, submission to authority and ignorance over knowledge and critical thinking. Many enforce strict rules of behavior that may be outmoded and inappropriate in today's world. They may play down the value of nonhuman life and promote the view that nature's purpose is to serve humanity. Believing in a close relationship with a personal God and exclusive knowledge of the truth may suppress our interest in solving earthly problems and respecting the beliefs of others. The history of religion includes a chronicle of conflict and warfare; religious fanatics have used religion to justify unspeakable acts of violence.

Although there are many arguments against theism, there are other religions that hold more plausible views. Deism affirms that a supreme power created the universe but no longer intervenes in the world, Pantheism claims that God represents all that there is and process theology asserts that God is evolving. They don't contradict current scientific knowledge. Some people adopt these non-theistic religions, whereas others prefer a more liberal form of religion or a form of spirituality devoid of religious affiliation. For some mystics God is all that exists, there is no separation between creator and creation. Others reach out to a transcendent presence which, although hidden from us, may be experienced in other ways.

The scientific worldview also has important shortcomings. It has helped us to better understand nature and improve our standard of living; however, there are still many gaps in our knowledge. In addition, scientific facts often don't assuage our existential fears. Nevertheless, the success of the application of the scientific method supports a degree of certainty about the tenets of science that is beyond reasonable doubt. Science represents the collective effort of many individuals that has resulted in the accumulation of a large body of information and numerous technical advances. Not all scientists embrace atheism, materialism and reductionism; some prefer methodological naturalism, whereas others are agnostic, religious naturalists or have spiritual, humanistic or religious affiliations. The most extreme religious position is that of fundamentalism. In contrast to those who follow traditional or liberal religions, fundamentalists believe that they are the only ones that have the truth. Traditional theists have a more moderate view. The theistic worldview, including Christian theism, provides answers to many existential questions but requires accepting dogmas that sometimes contradict reason and science. Many believers question some of these dogmas, the teachings of Scripture and have trouble explaining the problem of evil. Non-theists have more plausible religious views, they believe in a God that is nature, evolves with nature or exists outside nature and no longer interacts with the world. Liberal religions are open-minded, free from the constraints of dogmatism and authority. Many Unitarian Universalists don't believe in the divinity of Christ and are closely associated with humanism. They believe that every individual has value and should be free to search for meaning.

There are also many options for people without religious affiliation. Religious naturalists find meaning and value in nature but reject traditional supernaturalism. Secular humanists combine reason with an ethical dimension in order to make a better world. Secular humanists are hardly cold and insensitive rationalists; some of them, in addition to working for the good for humanity, place additional emphasis on caring for the planet and all life forms. Agnostics/seekers affirm that we cannot prove or disprove that God exists. Some people embrace methodological naturalism, a view that avoids supernatural explanations but doesn't assume that everything is physical. Others prefer ontological naturalism and positive atheism. Without claiming to have all the answers, there are many paths to choose from in dealing with life's most important questions, as suggested in Table 3..

TABLE 3. SPECTRUM OF RELIGIOUS AND NON-RELIGIOUS BELIEFS

| Religious fundamentalism |
| Traditional theism |
| Non-theistic religions |
| Liberal religions |
| Mysticism, spiritualism |
| Religious naturalism |
| Secular humanism |
| Agnosticism |
| Methodological naturalism |
| Ontological naturalism |

Chapter 14. A Rational Belief System

The choice between science and religion represents a false dilemma; we have enough scientific knowledge about the nature of reality to formulate a rational belief system. The proposed system is based on agnosticism, methodological naturalism, and humanism. This belief system doesn't need a canon of scriptures. It is supported by the scientific description of the universe and human nature. The religious scriptures should be accepted as historical documents that reflect the lives and beliefs of pre-modern people that lived during the "axial age." These books were written by these peoples in order to understand nature and to set rules for individual and group behavior. They are anonymous, come from several sources that sometimes contradict each other and are often biased by the interpretations of the author.

Agnosticism is a logical stance because we cannot prove the existence or the non-existence of a deity. This view doesn't exclude the possibility that an unknowable, ineffable deity exists; it is compatible with Deism, Pantheism and process theology. It questions, though, the tenets of traditional theism, especially Christian theism. Methodological naturalism is a worldview based on the study of nature using science and rationality, without embracing materialism or ruling out the existence of supernatural entities, like ontological naturalism does. Humanism is a philosophy of life that encourages living ethical lives that promote the greater good for the planet and all living creatures, including mankind. It supports justice and political systems that bring freedom and equality to everybody. Humanism has no creed other than the humanist manifestos, and rejects religious rituals and clerical hierarchies.

Agnosticism

The subject of agnosticism has not received the attention that it deserves. Agnosticism is often dismissed by the presumptuous argumentation between theists and atheists. The term was coined by Huxley; it derives from the Greek agnostos meaning "unknown" or "unknowable." He stated that it is wrong for a man to assert that he is certain of the truth without evidence that can justify this certainty. Russell defined agnostics as individuals who think that it is impossible to know the truth in matters concerning the existence of God or a future life. For Hume, agnosticism amounted to a total suspension of judgment about God's existence.

Agnostics have been criticized as being tentative fence-sitters, as irrational and as having a position indistinguishable from atheism. There are many versions of agnosticism based on the degree of probability of it being true that God exists. This characteristic separates agnosticism from theism and atheism. Some agnostics believe that it is highly improbable that God exists; in this case they are not too far removed from atheism. The modern versions of agnosticism agree with Mill who stated that the rational attitude toward the supernatural is that of skepticism. Some of these versions go further than Russell's agnosticism in that they investigate certain forms of theism that are more plausible than those considered in the past.

Methodological Naturalism

This is the view that many scientists embrace as the method of inquiry. It utilizes the scientific method and avoids the issues of materialism and the existence of supernatural entities. Science, though, is far from establishing with certainty the nature of reality. Scientists can't tell us why the universe exists and how it was created. They have trouble explaining the origin of life and the qualitative aspects of experience.

Scientists are aware that the nature of existence is a profound mystery. It is difficult for them to formulate a description of a reality that may be beyond our concepts and reasoning capacity. Some believe in the existence of God, a self-existing entity that created the universe. Other scientists, though, argue that the universe didn't have a beginning and that self-existence could be a property of the primordial form of matter/energy that had the capacity to evolve and become the source of all that there is. There is scientific evidence to support the idea that matter and energy are interchangeable and recent studies suggest that particles, such as the Higgs particle, may originate in a field.

Modern cosmology suggests that the big-bang is the origin of everything that exists. All the matter in the universe can be traced back to the big-bang and the death of stars; living and non-living things are both made of cosmic stardust. Because we don't have a workable quantum theory of gravity, scientists cannot describe the events before or at the time of the big-bang. Some scientists have embraced loop quantum gravity and others M-theory to explain the origin of the universe. Others postulate a cyclic universe having no beginning or end in time.

Scientists still don't understand the origin of life. They posit that life began at a point in time when matter became highly organized and acquired a limiting membrane, a source of energy and a mechanism of reproduction. Scientists believe that evolution by natural selection is the main mechanism responsible for the variety of species that exist in our planet. After millions of years of evolution, these life forms evolved into animals with complex brains and eventually into humans with highly developed minds and culture.

Evolution may have favored emotions and behaviors that originally were life-preserving, but now are a source of conflict. Exaggerated emotions may trump reason and become pathological: sadness becoming depression, fear leading to anxiety and anger turning into violence.

Most scientists hold that what we call mind/soul/spirit is the result of the activity of neural networks in the brain; nonetheless, it remains a mystery how neural processes in the brain create our experiences. Many neuroscientists believe that our experiences arise from emergent processes in neural networks; these processes may be physical but cannot be reduced to their individual components.

Humanism

The term "humanism" has many different meanings. It was originally used to describe a movement to revive classical Greek and Roman culture. During the Middle Ages Christian ethicists borrowed heavily from Greek philosophy, especially Aquinas who was a follower of Aristotle's. Others followed Cicero in ancient Rome and advocated free thinking, and at the same time, a sense of community and kindness toward humanity. The Renaissance gave birth to a broader conception of humanism, above and beyond the religious influence that had dominated the medieval mind. Secular humanism was definitively established during the Enlightenment. In the twentieth century the tenets of humanism were clearly articulated and were published in several manifestos that have been endorsed by both religious and secular humanists. The American Humanist Association is a secular group that affirms our responsibility to lead ethical lives of personal fulfillment that aspire to the greater good of humanity.

The question of how to live a good and ethical life has been difficult to answer. Socrates encouraged his disciples to think seriously about how they should live and Plato placed the idea of the Good as the ultimate goal. Aristotle felt that "practical wisdom" and good character are the best instruments to achieve the good life. During the Hellenistic period the Stoics advocated the need to remain indifferent towards things that we cannot control, and to rein in our passions, which we should be able to control. Epicureans recommended everyone to pursue pleasures and avoid pain; their pleasurable activities included intellectual and social activities not the sensual activities that we often associate with pleasure. During the Middle Ages, religion was the foundation of ethics, and the question of goodness was not separated from theology until the eighteenth century. Two of the ethical trends that emerged during this period, utilitarianism and deontology, are still under active discussion at the present time. As currently conceived humanism doesn't collapse into relativism or other philosophies of post-modernist persuasion.

Humanists believe that there are many worthwhile aspirations for human life and each individual has to find his own way of achieving them and to have a positive impact on the world. A life lived in this manner is said to be meaningful. Some of the common attributes of the good life include having close relationships with friends and family, pursuing creative activities according to personal abilities and being able to live authentically and exercise some degree of personal autonomy.

Many humanists question the suggestion that morality cannot exist without religion. They also may question the notion of free will — humans have a dual or Janus-faced nature consisting of primitive, mostly unconscious, emotions that originate in the limbic system and the capacity to reason, which is a function of the prefrontal cortex. Our cognition is sometimes fast and intuitive, other times slow and rational. We possess innate mechanisms responsible for cooperative behaviors, such as empathy and altruism, which are already present in other primates, but we may also exhibit competitive and aggressive behaviors that evolved in order to assure the survival of the individual and the species. Our social, political and economic history has been marked by the constant struggle between these opposing dispositions.

Because our dispositions to promote cooperation and social bonding do not always prevail, our ancestors introduced socially embedded ethical rules to deal with our selfish and competitive behaviors. Nevertheless, we often act irrationally, exhibit violent behaviors and threaten even our own survival (Feltham, 2007). Our record of violence, torture, terrorism and genocide is undeniable, we have accumulated enough nuclear, chemical and bacteriological weapons to destroy the world and we have failed to curb overpopulation, poverty, inequality, deforestation, mass extinctions and the production of inordinate amounts of greenhouse gases (Kirstin and Dow, 2006; Martin, 2006). The future course of humanity remains uncharted.

Humanists hope that the battle between reason and emotion will eventually be won by the former and that one day our pro-social dispositions for empathy and caring will trump our selfish and aggressive drives and emotions

Kurtz has recently drafted a neo-humanist statement with special emphasis in addressing the needs and goals of the planetary community. These include the need to strengthen transnational institutions capable to deal with global problems such as regional wars, climate change and the spread of infectious diseases. Cooperative action may be needed for prevent overpopulation, improve public sanitation and ameliorate poverty and hunger in the world. Of special concern is the problem of removing oppression from repressive social systems while respecting differing traditions. Different peoples have a right to define for themselves what is "right" in their society, while not jeopardizing the whole planet. We should resolve to realize the ancient dream that all human beings share certain needs and goals. We share a common abode, the planet Earth, and we should work together to realize the ideal of a true planetary community.

Neo-humanists aspire to be more inclusive and cooperate with both religious and non-religious people to solve common problems. They are skeptical of strict moral absolutes derived from religion and defend the separation of church and the state. Many are religious naturalists and have spiritual experiences when contemplating the wonders and mysteries of nature. Their ethical rules are not derived from theological absolutes; they originated in the light of modern inquiry. Progressive humanism supports not only the rule of law but the application of the principles of equality before the law and social justice. It also supports a green economy, the quest for new sources of clean energy and the avoidance of environmental degradation and pollution.

CHAPTER 15. CONCLUSION

In this book both religion and science are reviewed, along with many hybrid or open-ended worldviews, in order to understand how the different approaches seek to explain our world. No one has all the answers, but given the benefits of education and progress in science, in this era it makes sense to focus on a rational worldview based on the scientific evidence we have.

The wonders of nature will inspire and challenge our curious minds endlessly. As far as the existential problems we face as human beings, it is up to each person to find his place. There can be no "correct answers" to the perennial questions of how there came to be something rather than nothing and what is the origin of the universe. Human capabilities are limited. It is unclear how life forms originated from inanimate matter, although we will likely never stop seeking to know. For now, the best scientific theories posit mechanisms for the spontaneous formation of a proto-cell capable of reproducing and utilizing available energy sources. Examination of the tree of life suggests that all living organisms evolved from a primitive ancestor and that the theory of evolution by natural selection is the best explanation we have for the numerous species that exist in this planet.

Scientists find no evidence to support the notion of souls or to explain how they could interact with physical bodies. We now understand many of the mechanisms responsible for the function of the human brain, the most complex organ in the universe, but scientists still don't understand how the activity of neural networks creates the phenomenal experiences that we have.

Human nature is an expression of the results of the interaction of genes with the environment. Evolutionary psychology explores the theory that we are born with biological dispositions, formed during evolution, that are designed to assure the survival of the individual and the species. At the same time, the environment may influence our behavior and contributes to rewire our neural connections and even alter our genome.

Humans exhibit both reason and emotion; the former is localized in the prefrontal cortex and the latter in the limbic system. The prefrontal cortex is responsible for our advanced cognitive abilities and the reining in of our emotions; the limbic system is the source of our emotions and romantic ideals. As such, this is the region where any religious faith would be generated and entertained. Our rational brain helped us to develop an advanced civilization, and our emotions protect us and motivate our behavior. We feel that we have free will but our choices may be constrained by biological factors and unconscious desires. The qualities that we value as most human may not be based solely on psychological, cultural or economic factors; they may be biologically-based.

We experience conflicting emotions; we can be both cooperative and competitive. Our behavior is influenced by the balance between our altruistic tendencies and our self-interest. Cooperative behaviors may have their evolutionary origin in our disposition towards altruism, our capacity to show empathy towards other living creatures and our ability to bond to family and friends. Competitive behaviors arose in order to survive and reproduce. Humans tend to stereotype each other, perhaps to avoid potential predators, and seek status, resources and mates.

Humans face several existential issues; the two most important ones are death and meaning. To posit, though, that humans should have an explanation for everything, that we are so important that our lives should have a greater meaning beyond that which we can see, and to want an all-knowing and intentional being behind all this, seems to be asking rather much. Still, some nonreligious people find peace in the awareness that everything that exists is all made of the same "building blocks" and that all life forms are related: this can inspire a sense that we are interconnected with the rest of the universe.

We are fortunate to exist and to have the special senses that permit us to enjoy the sounds of romantic music, the beauty of sunsets and the smell of flowers. We are capable of acts of invention, creativity and heroism. We came into this world equipped with a magnificent body and mind forged by millions of years of evolution. Every cell in our body performs incredible tasks and a myriad of highly coordinated life processes. We possess a sophisticated brain, a complex organ capable of probing the deepest secrets of nature and creating sublime works of art and the wonders of modern science and technology. We may be the only, or at least the most, complex life form in the universe. We should use our unique capabilities, including our ability to reason, wisely.

REFERENCES

Barkow, Jerome, Leda Cosmides, John Tooby. *The Adapted Mind: Evolutionary Psychology and the Generation of Culture*. New York: Oxford University Press, 1992.

Bienenfield, David. *Psychodynamic Theory for Clinicians*. Philadelphia: Lippincott, Williams and Wilkins, 2006.

Carroll, Sean B. *Endless Forms Most Beautiful: The New Science of Evo Devo and the Making of the Animal Kingdom*. New York: Norton, 2005.

Chown, Marcus. *The Quantum Zoo: A Tourist's Guide to the Never-ending Universe*. Washington DC: Joseph Henry Press, 2006.

Clayton, Philip, Paul Davies. *The Re-emergence of Emergence: The Emergentist Hypothesis from Science to Religion*. Oxford; New York: Oxford University Press, 2006.

Damasio, Antonio. *The Feeling of What Happens: Body and Emotion in the Making of Consciousness*. San Diego: A Harvest book, Harcourt Inc., 1999.

Dawkins, Richard. *The Greatest Show on Earth: The Evidence for Evolution*. New York: Free Press, 2009.

Deamer, David. *First Life: Discovering the Connections Between Stars, Cells and How Life Began*. Berkeley: University of California Press, 2011.

DeDuve, Christian. *Singularities: Landmarks in the Pathways of Life*. Cambridge: Cambridge University Press, 2005.

Dennett, Daniel. *Breaking the Spell: Religion as a Natural Phenomenon*. New York: Viking, 2006.

DeSalle, Rob, Tattersall, Ian: T. *The Brain: Big Bangs, Behaviors and Beliefs.* New Haven: Yale University Press, 2012.

De Waal, Frans. *The Bonobo and the Atheist: The Search for Humanism among the Primates.* New York: WW Norton, 2013.

Diamond, Jared. *The Third Chimpanzee: The Evolution and Future of the Human Animal.* New York: Harper Perennial, 1992.

Fairbanks, Daniel J. *Evolving: the Human Effect and Why it Matters.* Amherst, New York: Prometheus Books, 2012.

Feltham, Colin. *What's Wrong with Us? The Anthropathology Thesis.* Chichester, West Sussex UK and Hoboken NJ: John Wiley and Sons Ltd., 2007.

Fiske, Susan T. *Envy Up, Scorn Down: How Status Divides Us.* New York: Russell Sage Foundation, 2011.

Flanagan, Owen. *The Problem of the Soul: Two Visions of Mind and How to Reconcile Them.* New York: Basic Books, 2003.

Greene, Brian. *The Hidden Reality: Parallel Universes and the Deep Laws of the Cosmos.* New York: Alfred A. Knopf, 2011.

Harris, James F. *The Ascent of Man: A Philosophy of Human Nature.* New Brunswick NJ: Transaction publishers, 2011.

Haack, Susan. *Defending Science-within Reason: Between Scientism and Cynicism.* Amherst, New York: Prometheus Books 2003

Hauser, Mark D. *Moral Minds: How Nature Designed our Universal Sense of Right and Wrong.* New York: Ecco, 2006.

Hawking, Stephen, Leonard Mlodinow. *The Grand Design.* New York: Bantam Books, 2010.

James, Scott M. *An Introduction to Evolutionary Ethics.* Chichester West Sussex UK; Malden MA: Wiley-Blackwell, 2011.

Kahneman, Daniel. *Thinking Fast and Slow.* New York: Farrar, Strauss and Giroux, 2011.

Kaneko, Kunihiko. *Life: an Introduction to Complex Systems Biology.* Berlin: Springer-Verlag, 2006.

Kazazian, Haig H. *Mobile DNA: Finding Treasure in Junk.* Upper Saddle River, New Jersey: FT press, 2011.

Kelly, Lynne. *The Skeptic Guide to the Paranormal.* New York: Thunder's Mouth press, 2004.

Kirstin, Dow, Thomas E. Dow. *The Atlas of Climate Change: Mapping the World's Greatest Challenge.* Berkeley: University of California Press, 2006.

Kurtz, Paul. *What is secular humanism?* Amherst, New York: Prometheus Books, 2007.

Lane, Nick. *Power, Sex, Suicide: Mitochondria and the Meaning of Life*. Oxford; New York: Oxford University Press, 2005.

Lieberman, Matthew D. Social, *Why our Brains are Wired to Connect*. New York: Crown Publishers, 2013.

Lang, Kenneth. *The Life and Death of Stars*. New York: Cambridge University Press, 2013.

Lenton, Tim, Andrew Watson. *Revolutions that Made the Earth*. Oxford; New York: Oxford University Press, 2011.

Lofton, John W. *Why I Became an Atheist*. Amherst, New York: Prometheus Books, 2008.

Martin, James. *The Meaning of the 21st Century: a Vital Blueprint for Ensuring our Future*. New York: Riverhead books, 2006.

Martin, Michael. *The Cambridge Companion to Atheism*. New York: Cambridge University Press, 2007.

Maynard Smith, John, Eors Szathmary. *The Major Transitions in Evolution*. Oxford; New York: W H Freeman Spektrum, 1995.

Meares, Russell. *A Dissociation Model of Borderline Personality Disorder*. New York: WW Norton, 2012.

Meyer, Thomas, Lewis P. Hinchman. *The Theory of Social Democracy*. Cambridge: Polity, 2007.

Minsky, Marvin. *The Emotion Machine: Commonsense Thinking, Artificial Intelligence, and the Future of the Human Mind*. New York: Simon and Schuster, 2006.

Mlodinow, Leonard. *The Drunkard's Walk: How Randomness Rules our Lifes*: New York, Pantheon Books, 2008.

Nielsen, Kai. *Ethics without God*. Amherst, New York: Prometheus Books, 1990

Niose, David. *Nonbeliever nation: the rise of secular Americans*. New York: Palgrave Macmillan, 2012.

Perrucci, Robert, Earl Wysong. *The New Class Society? Goodbye American Dream*. Lanham, MD: Rowman & Littlefield, 2008.

Pinker, Steven. *The Blank Slate: The Modern Denial of Human Nature*. New York: Viking, 2002.

Plante, Thomas G. *Mental Disorders of the New Millennium*. Westport, Conn: Praeger, 2006.

Riddley, Matt. *Genome: The Autobiography of a Species in 23 Chapters*. New York: Harper Collins, 1999.

Schweid, R. *Hereafter-Searching for Immortality.* New York: Thunder Mouth Press, 2006.

Shermer, Michael. *The Believing Brain: From Ghosts and Gods to Politics and Conspiracies - How We Construct Beliefs and Reinforce Them as Truths.* New York: Time Books, 2011.

Silverstein, Marshall L. *Disorders of the Self: A Personality Guided Approach.* Washington DC: American Psychological Association, 2006.

Smolin, Lee. *The Trouble with Physics: The Rise of String Theory, the fall of a Science, and What Comes Next.* Boston: Houghton Mifflin, 2006.

Stenger, Victor J. *The Comprehensible Cosmos: Where Do the Laws of Physics Come From?* Amherst, NY: Prometheus Books, 2003.

Swanson, Larry W. *Brain architecture: understanding the basic plan.* New York: Oxford University Press, 2012.

Trivers, Robert. *The Folly of Fools: the Logic of Deceit and Self-Deception in Human Life.* New York: Basic Books, 2011.

Tuschman, Avi. *Our Political Nature: The Evolutionary Origin of what Divides Us.* Amherst, New York: Prometheus Books, 2013.

Vilenkin, Alexander. *Many Worlds in One: The Search for Other Universes.* New York: Hill and Wang, 2006.

Wade, Nicholas. *Before the Dawn: Recovering the Lost History of Our Ancestors.* New York: Penguin Press, 2006.

Wilson, Edward. O. *Consilience: The Unity of Knowledge.* New York: Random House, 1998.

Wright, Robert. *The Moral Animal: Why We Are the Way We Are: The New Science of Evolutionary Psychology.* New York: Pantheon Books, 1994.